SUPERSTARS

SUPERSTARS

How Stellar Explosions Shape the Destiny of Our Universe

David H. Clark

McGraw-Hill Book Company

New York · St. Louis · San Francisco
Hamburg · Mexico · Toronto

1 2 3 4 5 6 7 8 9 FGRFGR 8 7 6 5 4

ISBN 0-07-011152-9

LIBRARY OF CONGRESS CATALOGING IN PUBLICATION DATA

Clark, David H.
Superstars: how stellar explosions shape the destiny of our universe.

1. Supernovae. I. Title.
QB841.C59 1984 523.8′446 83-14945
ISBN 0-07-011152-9

Book design by Grace Markman

To my family

CONTENTS

ACKNOWLEDGMENTS

Originality is the art of concealing your source.
—*Franklin P. Jones*

MUCH OF THE MATERIAL presented in this book is the result of past collaborative work: most notably the study of historical supernova records with Dr. Richard Stephenson, the study of radio supernova remnants with Dr. Jim Caswell, the study of possible climatic and biological effects of supernovae with Prof. William McCrea, the study of optical supernova remnants with Dr. Paul Murdin, and the study of X-ray supernova remnants with Dr. Ian Tuohy. It is a pleasure to acknowledge with sincere thanks their contributions to my understanding of the superstars. I am particularly indebted to Prof. Bernie Mills, who first introduced me to supernova research via observations of their radio remnants almost a decade ago.

In the belief that a book of this type is not the place to list detailed references, I have adopted the technique throughout the main text of referring by name only to those astronomers no longer living or not now actively working in the field of supernova research. It is often difficult to assign priority of claim to recent discoveries and advances in interpretation—history requires time to make such judgments. Specialists will know the identity of those who have made significant observational contributions and recognize the source of many of the ideas presented, but such details are of little concern to others. I hope I shall not have offended anyone by failing to acknowledge their claim to original work. I wish, however, to mention several individuals working in supernova research, in addition to those listed above, whom I have enjoyed stimulating conversations with in recent years and whose published papers have greatly influenced my own work: they are, in alphabetical order, Dave Arnett, Dave Branch, Roger Chevalier, Stirling Colgate, John Danziger, Mike Dopita, Miller Goss, Dave Helfand, Bob Kirshner, Dave Schramm, Fred Seward, Gustav Tammann, Sidney van den Bergh, Kurt Weiler, Craig Wheeler, Frank Winkler, and Roger Wood. I hope to do justice to the many ideas with which they have reshaped supernova research in recent years.

Samuel Johnson noted, "What is written without effort is in general read without pleasure." This text has *not* been written without effort. However, transforming the writings of a novice author into a book which I hope might be read with pleasure has required a great deal of editorial assistance—I wish to express my sincere thanks to my publishers and my agent, Felicia Eth of Writers House. Finally, I wish to thank my family for all the encouragement and help they have given me in this endeavor, particularly my wife Suzanne for her work preparing the manuscript and for all her love and support.

DAVID H. CLARK

INTRODUCTION

> Experience has shown, and a true philosophy will always show, that a vast, perhaps the larger portion of truth arises from the seemingly irrelevant.
>
> —*Edgar Allan Poe*

MODERN SCIENCE HAS REVEALED that on the cosmic scale, planet Earth is a mere speck of celestial sand swept by the tides of universal change. Our solar system is only one among many, and that our galaxy is one of myriads in a universe experiencing violent upheaval and change. We have come to understand that what happens outside the solar system has profoundly influenced the nature and indeed the very existence of our own planet.

Among the spectacular phenomena influencing the evolving universe, exploding stars, known to astronomers as supernovae, have a special place. Supernovae have had a profound impact on humankind throughout history: they have been held in great awe and recorded and studied in considerable detail. But the influence of these "superstars" has not been limited merely to those privileged to witness such celestial pyrotechnics. Modern investigations have suggested that our Sun and its planetary system may owe their birth to a supernova, and that subsequent nearby supernovae may have profoundly influenced the biological and climatic evolution of planet Earth.

Could a supernova have initiated the formation of the solar system? Were certain of the heavy elements, such as gold and uranium, formed in supernovae? Could a supernova have caused an ice age? Could a supernova have killed the dinosaurs? Was the Star of Bethlehem a supernova? Could the witness of spectacular supernovae have stirred the human imagination to a new awareness of the wonders of the heavens? Do supernovae produce the enigmatic "pulsars"? Research into the nature of the superstars, with evidence for and against these tantalizing possibilities, is the subject of this book.

The story of the superstars brings together a number of fascinating and what at first sight must appear to be totally unrelated research disciplines. It would be difficult to imagine any other concatenation of circumstances calling on the investigative skills of historians, Sinologists, archaeologists, astronomers, physicists, engineers, climatologists, paleobiologists, and a host of others. While many of the world's great telescopes study the debris or remnants of supernovae, information of comparable value may be gleaned from ancient manuscripts describing early observations of stellar explosions. The application of data from giant telescopes, symbols of a technological age through which we have extended and improved our senses, and from historical literature recording the resoluteness and perceptiveness of the stargazers of ancient times armed with no more than the power of the unaided eye brings a remarkable and unexpected continuity to the study of one of nature's greatest spectacles.

Our present cosmic quest therefore involves the contrasts of historical literature and the most sophisticated instrumentation technology is able to offer. Modern astronomers are not restricted to observing the heavens in the visible light to which the human eye is sensitive: massive radio telescopes now provide them with "radio spectacles" through which they can view the heavens. The picture of a universe undergoing violent upheaval and change is dramatically different from the apparently quiet universe men and women have observed since antiquity as they looked with unaided vision out into the night sky. Futuristic spacecraft now probe the depths of space for cosmic radiations that could never be observed from the Earth's surface, their discoveries adding to the multicolored panorama of the universe modern astronomers continue to paint. The cosmic canvas may never be completed, however, since the true complexity of the universe may be beyond human understanding. Nevertheless, our attempts to understand the nature of our celestial heritage must surely represent one of the great human challenges, comparable to the most courageous exploits of adventurers throughout the ages who sought to explore the unknown. It is our capacity to understand and utilize our terrestrial environment that has set us apart from other earthly creatures. Perhaps it will be our capacity to partially understand our universal environment, as we seek our ultimate roots among the stars, that may eventually set us apart from other celestial civilizations.

The astronomer's picture of the universe differs from that put forward by, for example, poets, artists, theologians, or philosophers. None of these differing views of the universe is necessarily less valid than any other; they must be considered complementary. We will be concerned in this book principally with the universe revealed by astronomical research, a picture no less beautiful than that presented by any poet or artist, nor less pleasing than that proposed by any philosopher or theologian. What we will be discussing will be currently held scientific beliefs based on centuries of observation and evolving astronomical ideas. The scientific facts presented, therefore, must be seen to be the ideas that are found to be acceptable to the majority of scientists at the present time, rather than ultimate truths. Astronomical monographs and journals record the state of modern scientists' understanding of the universe, just as in their time did the cave paintings of primitive cultures. And as surely as the latter were superseded, so our current astronomical ideas of the universe can be expected to change. They are no less remarkable because of this limitation and remain testimony to the imagination and investigative powers of the human race and its desire to understand its universal environment. The story that follows is about the role of the superstars in our continuing cosmic adventure.

1

STARS IN CATASTROPHE

> There is nothing new except that which has become antiquated.
>
> —*Mlle. Bertin*

MOST STARS ARE EXTREMELY STABLE, as a casual glance at the night sky will reveal. Individual stars in the heavens look much the same, night after night, year after year. Their twinkling is not intrinsic but is an effect produced by scattering of light in the disturbed night air. Some stars do show a small variation in intensity, with the period between times of peak brightness ranging from fractions of a second, in the case of so-called pulsars, to tens or even hundreds of years. Such variations are usually indiscernible to the naked eye, and for the most part it would seem stars do produce a remarkably steady output of light over millions of years. In marked contrast are the novae and supernovae—stars that spontaneously explode with a spectacular and rapid increase in brightness over a few days, so that at maximum they may rival the brightest stars in the heavens, then gradually fade into insignificance in the course of anything from weeks to years. The term *nova* means literally "a new star"; it is in fact a misnomer.

It is one of the great paradoxes of astronomy that what were long thought to be new or super-new stars—the novae and supernovae—are really spectacular and catastrophic outbursts from very old stars. A supernova is now recognized as the violent end to the evolution of a certain type of star. The energy released in this cataclysmic act of stellar suicide is almost beyond comprehension; it can be estimated to be equivalent to the simultaneous energy release from about 10 billion, billion, billion 10-megatonne hydrogen bombs. (One "billion" equals a thousand million.) The outer envelope of the exploding star is ejected, broken up like shrapnel from a bomb, with velocities of thousands of kilometers per second. It is preceded by a shock wave expanding from the site of the holocaust. And, at least in some cases, the core of the star implodes, leaving a rapidly rotating collapsed stellar remnant. At its brightest a supernova may be intrinsically several billion times brighter than the Sun, so that if a supernova were ever to occur at a distance from the Earth of less than about 100 trillion kilometers (one "trillion" equals a thousand billion), it would turn night into day. Before looking at why certain stars undergo such catastrophic self-destruction, we will digress so as to set the cosmic scene for the drama of the birth, life, and death of stars.

Our understanding of physical dimensions such as distance, mass, and time is limited by our everyday experiences, although this understanding has changed dramatically during the present century. Medieval astronomers voyaged the thousands of kilometers around the globe in time scales of years; modern adventurers have completed the three-quarter-million kilometer round-trip journey to the Moon within days. Just over eighty years ago the first motor-driven flight carried one man a mere 40 meters. The Wright Brothers' first flight at Kitty Hawk could have been completed within the economy-class section of a modern Boeing 747 jumbo jet, which now carries over 400 passengers at a time across the continents in just a few hours. Yet this newly acquired global freedom and appreciation of terrestrial distances ill-prepares us for accepting the enormous distances of the cosmos. A journey to Pluto, our solar system's outer planet, in a supersonic Concorde jet aircraft, would take 300 years. A Concorde trip to the nearest star beyond the solar system, Proxima Centauri, would take over 2 million years! On such a scale, it makes little sense to measure distances in kilometers. Instead, astronomers define a unit based on the distance light travels through free space in a year—called a light-year. Light travels at the incredible speed of 300,000 kilometers each second. Thus a pulse of light could circumnavigate the globe eight times in a second—or complete the journey to the Moon in just a little longer. The light that reaches us from the Sun takes eight minutes to cross the 150 million kilometers. The nearest star is four light-years distant. If we were able to travel at close to the speed of light—that is to say, almost half a million times faster than the Concorde's "snail's pace"—it would still take us almost a decade to complete the round-trip journey to Proxima Centauri. One light-year is equivalent to 9 trillion 460 billion kilometers.

In establishing the distance of the nearest star to the solar system, we still have no real appreciation of the enormousness of the universe. Stars are not uniformly scattered throughout space but conglomerate in galaxies containing many billions of stars. Galaxies themselves accumulate into clusters. Our Sun is just one of an estimated 100 billion stars within our own galaxy, which is called the Milky Way. The Milky Way is discus shaped, a full 100,000 light-years across at its widest, with our Sun occupying a rather insignificant site closer to its periphery than its center. If we could look down at our galaxy from above, the bright stars of the Milky Way would appear to be concentrated within intertwined spiral arms.

While our place within the Milky Way may seem insignificant, the place of our galaxy within the universe is equally so. The observable universe is now believed to contain at least 10 billion galaxies, clustering together in the hundreds or thousands but still spaced from one another by many millions of light-years. The most distant galaxies observable with giant telescopes are 10 billion light-years distant!

Observing the cosmos through a large telescope means looking not only deep into space but also back in time. Because of the finite speed of light, the telescope acts as a time machine, revealing the nearby stars as they were a few years or tens of years ago and the more distant stars within the Milky Way as they were hundreds or thousands of years ago when the light now reaching the Earth commenced its cosmic journey. The nearby galaxies appear as they were millions of years ago and the more distant galaxies as they were hundreds to thousands of million years ago. Few of the objects observed exist at this instant, at least in the form we presently see them! Thus the history of the universe turns like a cosmic kaleidoscope for Earthbound astronomers. Astronomers can study stars and galaxies at various stages of their evolution—nascent stars, young stars, middle-aged stars, old stars, dying and dead stars, young galaxies, interacting galaxies, galaxies in formation, and galaxies being torn apart! The universe reveals itself as a site of unfolding drama, as stars and star systems, often violently, are born and die.

Further adjustment to our terrestrial way of thinking is required to appreciate the mass and time scales of the universe. We measure the mass of objects of common experience in terms of a convenient standard, the kilogram. Thus, for example, an adult human may typically have a mass up to about 80 kilograms. The mass of planet Earth, however, is 6 trillion trillion kilograms, while the Sun is some 300,000 times more massive. On the astronomical scale, we tend to use the mass of our Sun as an appropriate standard. Hence the mass of a star could be described as being so many solar masses. The mass of the Milky Way is believed to be on the order of 100 billion solar masses—and the mass of the universe is certainly greater (perhaps very much greater) than 300 billion billion solar masses!

No less of a challenge to the human imagination are the time scales involved in describing astronomical phenomena. Earthbound events are conveniently defined by the sidereal day, the time required for the Earth to revolve once about its axis measured with respect to the fixed stars, and the sidereal year, the time it takes the Earth to complete one orbit about the Sun, again measured relative to the fixed stars. The Sun and other stars orbit around the center of the Milky Way, like a gigantic cosmic Catherine Wheel. At the Sun's distance, a star takes over 200 million years to complete a single revolution of the Galactic center—an interval known as the Galactic year. Just as a human lifetime typically spans several tens of terrestrial years, so stars of similar mass to the Sun have lifetimes spanning several tens of Galactic years. Strangely, the more massive a star, the shorter its life expectancy! The Sun has survived for about 20 Galactic years since its birth and is expected to survive in its present state for at least a similar interval. But what of the age of the universe itself? This remains somewhat uncertain, but scientists now believe that the most recent epoch of creation, heralded by what is referred to as the Big Bang, was between 10 and 20 billion years ago! Astronomers must obviously "think big" when describing the nature of the universe, but in studying the evolution of stars they must also "think small," since it is processes on the atomic scale that eventually determine the fate of a star.

Let us now bring up the curtain on the first act of our drama. Stars are formed from the gas and dust that permeate the void of space. While it may not be apparent when looking out into the night sky, the space between the stars is not empty but contains an extremely rarefied gas, usually referred to as the interstellar medium. Its dominant components are hydrogen and helium, but the gas mixture is far from pure, containing tiny dust particles that obscure from view the more distant stars in the Milky Way. Nor are the gas and dust uniformly distributed around the stars. Rather, they are concentrated in condensations or clouds. Such clouds will include the debris of stellar explosions from a previous epoch. Occasionally, for reasons we will discuss later, a cloud collapses to form a star or multiple-star system. Once the collapse has started, it continues under the action of gravitational forces. Gravity, the all-pervading, universal attractive force acting between any two bodies, holds us and other objects to the Earth. It also holds the Moon in orbit around the Earth, the Earth and other planets in orbit about the Sun, and the Sun and other stars in orbit about the center of the Milky Way. On the universal scale, the action of gravity dominates other natural forces, molding planets and stars, star clusters, galaxies, and galaxy clusters.

As a nascent star forms from a cloud of gas, mainly hydrogen, and collapses under the action of its own gravity, the temperature, pressure, and density in the interior of the cloud increase dramatically. The gas pressure may reach 10 billion times atmospheric pressure—an awesome figure when we consider that the highest steady pressure that can be attained in the laboratory is just a few million "atmospheres"! The temperature at the center of the collapsing gas cloud may reach more than 10 million degrees. This temperature is enormous in the context of everyday experience; water boils at 100 degrees celsius, the temperature in a furnace may reach a few thousand degrees, and the temperature of the visible surface of the Sun is about 6,000 degrees. If the surface temperature of the Sun matched that at its center (about 15 million degrees), the inner planets of the solar system (including the Earth) would be vaporized!

At the extreme internal temperatures and densities of the Sun and other stars, certain nuclear burning reactions referred to as nuclear fusion can take place. Normal chemical burning processes are hopelessly inadequate to keep a star shining; for example, if the Sun were composed entirely of coal, with adequate oxygen to sustain combustion, it would burn away in just a few thousand years.

To present a detailed discussion here of nuclear reactions might be considered an unnecessary digression by some readers already familiar with basic physics or by those who do not wish to probe too deeply the physics of supernovae. Therefore, such material is included in the Appendixes. The structure of atoms is covered in Appendix 1, while the nature of nuclear reactions themselves is covered in Appendix 2. It will suffice here to note that all matter is made up of basic building blocks called atoms. The bulk of the mass of an atom is concentrated in a small, dense central nucleus around which orbit minute electrons. If an atom loses some of its electrons, it is said to be ionized. The nucleus is made up of particles called protons and neutrons. Nuclear fusion reactions involve the merging of atomic nuclei, with energy being released in the process.

In a newly formed star, hydrogen nuclei are fused to form the heavier helium nuclei, with the release of energy. The liberation of this so-called thermonuclear energy sufficiently increases the pressure in the mass of gaseous material making up a newly forming star to halt the gravitational contraction producing it. A star is born, and the curtain comes down on act one. As act two begins, the young star settles down to the relatively stable state in which it spends most of its active life. During this long period of stability, referred to as the main sequence of stellar evolution, the star's self-gravity acting inward is balanced by the pressure pushing matter out. Energy generated in the star's interior is transported to the stellar surface, exactly compensating for energy radiated into space. This delicate balancing act is maintained at the expense of a loss of nuclear fuel. In a star like our Sun, about 655 million tonnes (a tonne is one metric ton) of hydrogen are transformed into about 650 million tonnes of helium each second. The lost mass of about 5 million tonnes per second is converted to energy. Even with mass being destroyed at this rate, there is enough hydrogen in the Sun to maintain a steady output of energy for more than 10 billion years. Over this period, the total mass of the Sun will decrease by less than one tenth of one percent.

And so it is with all stars. The loss of mass and the generation of thermonuclear energy provides the answer to the question that has sparked the curiosity of people over the millennia: What makes the Sun and stars shine? In fact, the secret energy source utilized with potentially catastrophic consequences in the building of thermonuclear (hydrogen) bombs is the very energy source successfully harnessed and controlled in the central furnaces of the stars. It is one of the great ironies of science that the energy from the Sun that sustains life on planet Earth is identical in origin to that held in store in the nuclear arsenals of the world's superpowers. We have yet to learn to control the nuclear fusion process for peaceful applications.

In the extreme temperatures of stellar interiors, most electrons are stripped from their parent atoms and move rapidly along random paths between the tightly packed atomic nuclei. Sir Arthur Eddington presented the following graphic description of conditions on the atomic scale in the gaseous globe of a star:

> Try to picture the tumult! Dishevelled atoms tear along at 50 miles a second with only a few tatters left of the elaborate cloaks of electrons torn from them in the scrimmage. The lost electrons are speeding a hundred times faster to find new resting places. Look out! There is nearly a collision as an electron approaches an atomic nucleus; but putting on speed it sweeps around it in a sharp curve. A thousand narrow shaves happen to the electron in [10 billionths of a second]; sometimes there is a side-slip at the curve, but the electron still goes on with increased or decreased energy. Then comes a worse slip than usual; the electron is fairly caught and attached to the atom, and its career of freedom is at an end. But only for an instant. Barely has the atom arranged the new scalp on its girdle when a quantum of [radiation] runs into it. With a great explosion the electron is off again for further adventures. Elsewhere two of the atoms are meeting full tilt and rebounding, with further disaster to their scanty remains of vesture. As we watch the scene we ask ourselves, "Can this be the stately drama of stellar evolution?"

The constraints imposed on the behavior of electrons in a stellar interior are discussed in Appendix 3.

Although the nuclear fuel reserves of a star are enormous, they are not unlimited. When the core hydrogen is expended, gravity again takes control. As the core starts to contract, it raises internal temperature to the approximately 200 million degrees needed to start the nuclear burning of the helium ash left over from the earlier hydrogen transmutation. Increased temperature means increased pressure, so the star eventually stabilizes again. Helium nuclei fuse to form the heavier elements carbon and oxygen, accompanied now by hydrogen burning in the shell surrounding the central core. The star has increased in size and luminosity, and although the core temperature has increased, the inflated outer envelope of the star has cooled; the star has become what is called a red giant. What happens to the star now depends on its mass. A star with mass less than about 1.4 times the mass of our Sun (a mass referred to as the Chandrasekhar limit) never achieves high enough core temperature to sustain nuclear burning reactions beyond the formation of carbon and oxygen and undergoes a relatively controlled collapse to a state of very high density, forming a so-called white dwarf. Some stars with initial masses greater than 1.4 solar masses may also achieve this peaceful demise if they shed, as some stars do, large amounts of mass during their lifetimes. In forming a white dwarf, a star the size of the Sun contracts to a sphere about the size of the Earth, and at this volume a concentration of solar mass produces a density of several tonnes per cubic centimeter. In a white dwarf, electrons and the nuclei of the elements created from the various stages of nuclear burning will be tightly packed; it is the pressure exerted by the constantly moving, closely packed electrons that eventually counters the chaos gravity has attempted to produce. The form of the electron pressure is considered in Appendix 4. It is believed that 999 out of every 1,000 stars in the universe will become white dwarfs, the small, compact receptacles of stellar nuclear garbage. White dwarfs cool over an indefinite period, eventually becoming unobservable as cold, lifeless black dwarfs. The evolution of a solar-type star is depicted in Figure 1.

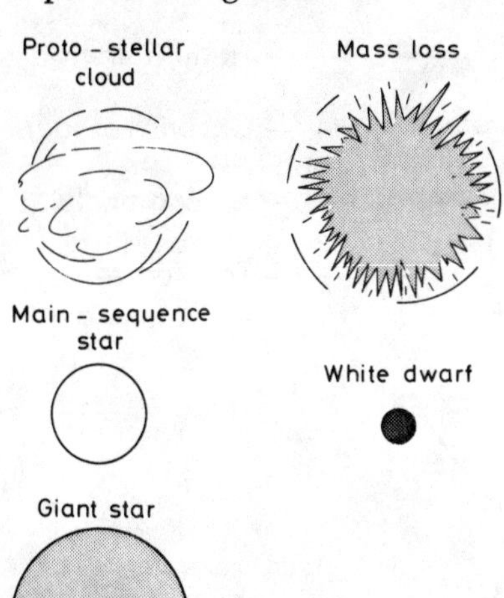

Figure 1. Schematic representation of the evolution of a Sun-like star.

But what of that one in a thousand stars which does not evolve into a white dwarf? For stars larger than a few solar masses, nuclear burning after all the helium in the core is expended leads to the fusion of successively heavier elements all the way to iron, when no further energy can be extracted. Several stages of nuclear burning can occur simultaneously in a single massive star (see Figure 2), with the central core reaching the iron end point while the outer portion of the stellar envelope is still burning hydrogen, with intervening zones at intermediate stages. Later in its evolution, the star expands greatly to become a supergiant. The star is now running into an energy conservation problem with catastrophic consequences. Since four hydrogen nuclei are required to create each new helium nucleus, there are only a quarter as many helium nuclei as there were original hydrogen nuclei available to un-

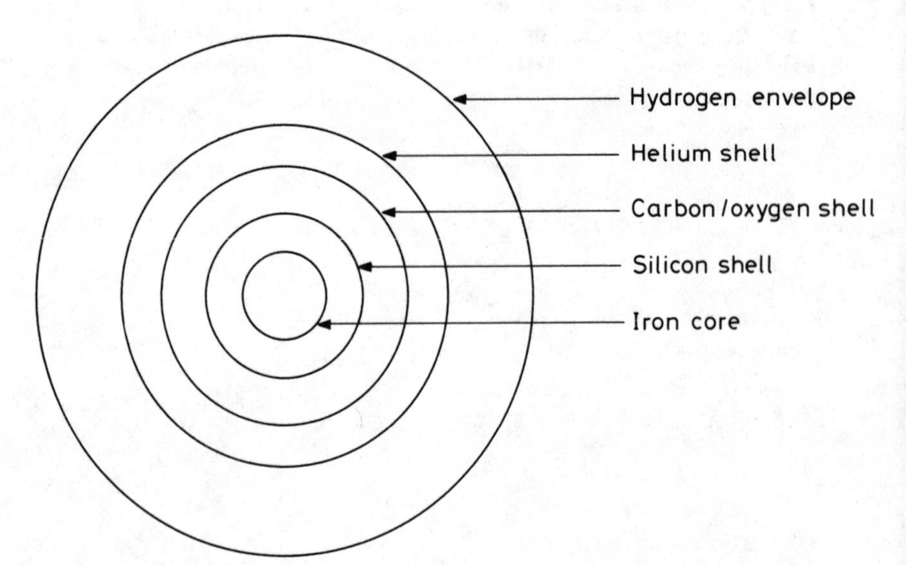

Figure 2. The structure of a massive star at an advanced stage of evolution, showing products of nuclear fusion in its shells.

dergo fusion. The number of reacting nuclei decreases similarly at each successive stage of nuclear burning. To make matters worse, the energy yield per fusion also decreases at each stage. The result is that after spending 90 percent of its life in the long, stable, hydrogen-burning phase, the star squanders its remaining fuel reserves at an ever-increasing rate. In the last stages of stellar evolution, a significant fraction of the energy released in the nuclear reactions in the core is in the form of strange, will-o-the-wisp particles called neutrinos, which can easily escape through the outer fabric of the star. Advanced warning of a star approaching its demise in a supernova event would thus be a dramatic rise in its neutrino emission. Perhaps in the future, telescopes will be sensitive enough to reveal neutrinos and thus permit scientists to identify stars approaching destruction.

A 25-solar-mass star is believed to spend 7 million years burning hydrogen to helium with core density about five times that of water, 500,000 years burning helium to carbon and oxygen with a 150-fold increase in core density, 600 years burning carbon to neon at a density 200,000 times that of water, just one year burning neon, a mere two months burning oxygen to silicon, by which time the density is 10 million times that of water, and less than a day burning its final silicon reserves to nickel, cobalt, and iron. The end is nigh; we have reached the final act. Drained of its energy reserves, gravitational collapse of the core of a massive star is inevitable. Once initiated, the catastrophic collapse takes place in a fraction of a second. In the extreme conditions of core collapse, electrons and protons merge to become neutrons. It may be possible for the pressure exerted by the tightly packed neutrons to take control and stop the core from contracting further (see Appendix 5). The central core of a 10-solar-mass star will have collapsed to a sphere a mere 15 kilometers in diameter consisting almost entirely of neutrons—forming a so-called neutron star. Even a mere pebble of neutron star would weigh 10 million tonnes. Neutron stars may be observable as pulsars, emitting pulses of radiation at intervals of a fraction of a second.

When neutron pressure is insufficient to halt the contraction of the core, very massive stars collapse to become black holes (see Appendix 6).

In describing the formation of neutron stars, we discussed only the behavior of the stellar core. But what of the rest of the star? The stellar envelope, still containing elements lighter than iron from intermediate stages of nuclear burning, collapses behind the core. As a result of its large energy at infall, the core overshoots nuclear density and bounces back. But this bounce is highly damped, so that the core rebounds just once, then stops. The bounce drives a shock wave of pressure outwards. A shock wave is produced whenever a disturbance in a gas cannot be dispersed fast enough, setting up a pressure enhancement—a common example is the shock produced ahead of an aircraft traveling at supersonic speed. Whether the outward-moving shock created by core bounce in a dying star can reach the collapsing hydrogen-rich outer envelope or will merely stall and die depends on a number of factors. Some shocks gain sufficient energy to blast the envelope of the progenitor star out into space. The old, dying star has been torn apart; the result is a super-new star—a supernova. A supernova which has its origins in the collapse of massive star is known as a Type II supernova and is depicted schematically in Figure 3. And so the final curtain falls on the drama of the life of a massive star.

In contrast to a supernova, which is a one-time event heralding the final destruction of a star, a nova is believed to be

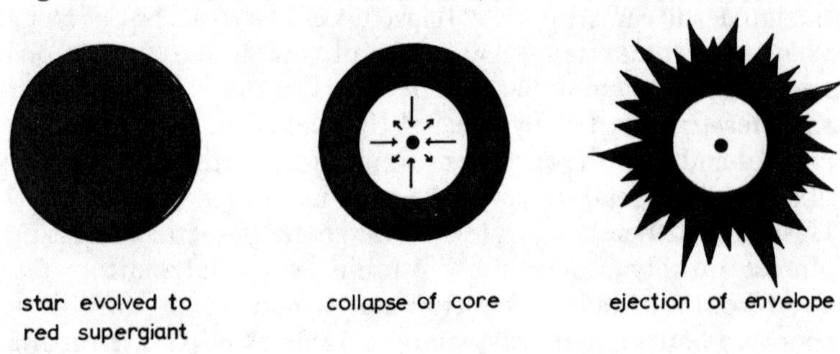

Figure 3. Schematic representation of the self-destruction of a massive star in a Type II supernova event.

merely a hiccup in the normal evolutionary path of certain stars and may happen repeatedly in intervals of up to several thousand years. A nova explosion releases less than one ten-thousandth of the energy of a supernova. In each nova explosion only a very small amount of material is believed to be ejected, and some of it may, under gravity, eventually fall back onto the star. The most likely explanation for these sporadic outbursts is based on the now convincing observational evidence that a prenova is a close binary system, in which two stars orbit around one another and are close enough for material to be transferred between them at certain times during the normal course of their evolution. Any close binary system containing a white dwarf and a normal companion star is a potential nova candidate. During an expansion phase of the companion star, material, mainly hydrogen, from its outer regions may come under the gravitational action of the white dwarf and be sucked from the star into the atmosphere of the white dwarf. Material is transferred, not directly but by means of a ring—an accretion disk—deep within the gravitational field of the white dwarf. There, under the action of the white dwarf's strong gravitational forces, the material is compressed and heated to the ignition temperature for nuclear burning. From this moment, thermonuclear runaway could produce the explosive ejection of matter and the radiation that are seen as the nova event.

A nova explosion resembles a supernova, but on a much smaller scale. A typical nova lasts days or at most weeks, whereas a supernova can last for months or years. The expanding shell of gas from a nova explosion dissipates in a few decades, leaving the binary system more or less in its original condition. The exchange of material in a close binary system is shown schematically in Figure 4.

The eventual fate of a white dwarf in a close binary system depends on the composition of the white dwarf and the rate at which material is transferred to it from the companion star. However, under certain conditions, which we will discuss later, as mass transfer pushes the white dwarf toward the Chandrasekhar limit, explosive nuclear reactions that can

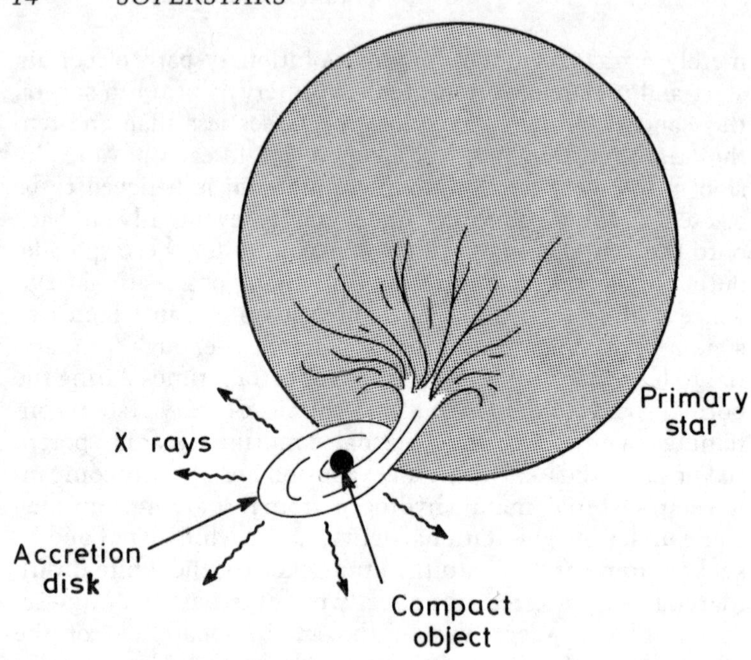

Figure 4. Mass exchange in a close binary system.

blast the white dwarf apart may be initiated. Thus is heralded the birth of a supernova event referred to as Type I.

Novae are much more frequent than supernovae, and for a short time a nearby nova can shine among the brightest of stars in the sky. It is the supernovae, however, that have proved to be of particular interest to many astronomers. Not only do they represent the most spectacular of stellar events, but the remnants and ejecta are among the most unusual and exciting of astronomical objects and phenomena. Supernova explosions are believed to be responsible for the production of high-velocity runaway stars hurtling through the Galaxy at speeds approaching a million miles an hour. Furthermore, the mysterious cosmic rays—energetic particles continually bombarding our planet—are believed by some to have originated in supernovae, and some supernovae are the likely progenitors of pulsars. Some of the most spectacular nebulosities in the heavens, such as the mischievous Crab, the beautiful Veil nebula, or the giant filaments of Vela, are the debris of ancient supernovae. The remnants may also emit radio waves and X rays. Such is the magic of the superstars.

2

RECORDS, RELICS, AND REMNANTS OF ANCIENT SUPERSTARS

> Look no more for another bright star
> by day in the empty sky
> more warming than the Sun . . .
> —*Pindar, "First Olympian ode"*

THERE HAVE BEEN MANY spectacular stellar outbursts recorded throughout history. Most pre-Renaissance observations of "new stars" were made in the Far East—China, Japan, and Korea. By contrast, there seems to have been very little interest in such phenomena in ancient and medieval Europe and in Arab lands.

The lack of interest in the appearance of new stars in the medieval Occident must have been partly due to the widespread influence of Aristotelian dogma. The great Greek philosopher Aristotle, in the fourth century B.C., formulated a picture of the universe where everything sublunary—that is, below the Moon—was made from four basic elements: fire, air, water, and earth. The superlunary regions, above the Moon, included a more perfect type of matter called ether. Earthly matter was deemed corruptible, subject to change and decay, but the heavens were perfect and immutable. Thus, transient phenomena such as comets, meteors, and aurorae were believed to be of atmospheric origin. (We now know, of course, that aurorae are indeed atmospheric phenomena and that meteors are fragments of interplanetary debris burning up on entering the Earth's atmosphere, but comets are true members of the interplanetary population.) The doctrine of heavenly perfectibility, which influenced Christian theology and Western scientific development through the Middle Ages until the Renaissance, could not admit the concept of a new star. However, perhaps a more important reason for the scarcity of new-star sightings from pre-Renaissance Europe was the Europeans' lack of interest in the heavens and their subsequent inability to recognize anything but the most spectacular of new starlike objects. In the year 1006 a new star blazed forth, so brilliant that even the Europeans could not ignore it. With the exception of the records of this new star, the paucity of ancient and medieval European records of new stars is disappointing in the extreme.

No new star was recorded in Europe between the year 1006 and the Renaissance, although two of noteworthy brightness and long duration were recorded in the Orient in 1054 and 1181. By the sixteenth century, however, an era of enlightenment had begun in Europe. Aristotelian dogma was being openly questioned by influential scientific voices. The appearance of two spectacular new stars in 1572 and 1604, and their detailed observation, helped finally to destroy the doctrine of celestial perfection. A heaven that produced transient stellar phenomena was clearly far from immutable. It was fortunate for the cause of science that astronomers of the caliber of the great Tycho Brahe (after whom the 1572 new star was named) and his pupil Johannes Kepler (for whom the new star of 1604 was named) were able to leave records of new stars more detailed than those for any previous stellar event. The tragedy was that their work preceded the advent of the astronomical use of the telescope (in 1609) by but a few years.

The situation in the ancient Orient was very different. From very early times each Chinese ruler was believed to be the Son of Heaven, appointed on a mandate from Heaven and able to rule only so long as he fulfilled Heaven's wishes. To the ancient Chinese, Heaven was more than merely the void of space and was believed to have a controlling influence on the destinies of men. Any departure by the ruler from the path of virtue would be expected to be signaled by signs in the heavens—the appearance of comets, meteors, and new stars. A rigid system of political astrology was thus established in China and copied, after the tenth century, in Korea. Professional astronomers/astrologers were appointed to maintain a constant watch of the sky and report and interpret unusual events, many of which were recorded in the official histories of the various dynasties. In addition to their astrological value, certain astronomical observations were necessary to provide a reliable calendar—an important requirement for a civilization dependent on agriculture. The custom developed that acceptance of the calendar promulgated by an emperor represented submission on the part of his subjects. A ruler's hold on his people depended as much on the calendar's accuracy as on military success abroad and political success at home. By 440 B.C. the year had been correctly calculated to have 365¼ days. Other moderately accurate data, such as information on planetary motions, had been collected by about 300 B.C. The comet later to be named after Edmund Halley was first observed in ancient China in 240 B.C. Observations continued in an almost unbroken succession, although the pre-Renaissance observers were unaware that they were charting repeat visits of the same celestial interloper. Such interest in heavenly bodies inevitably led to speculation about the nature of the universe. A philosophy of infinite empty space developed in the East, more than a millennium ahead of comparable Western thinking, as writings from the second century show: "The heavens are empty and void of substance. When we look up at it we can see that it is immensely high and far away, having no bounds. The Sun, the Moon, and the company of stars float in the empty space, moving or standing still. All are condensed vapour . . ."

Remarkably detailed astronomical records exist from China, at least since about 200 B.C. Very few records remain from earlier times because of the widespread burning of books in 213 B.C., instigated by Ch'in Shih Huang Ti, who had unified China in 221 B.C. and become its first emperor. The purpose of this literary holocaust was to eradicate all memory of the warring states that had vied with the now-ruling state of Ch'in. To ensure that the histories were not rewritten, the emperor also ordered that 400 of the country's top academics be buried alive. The systematic burning and sacking of the Ch'in capital that accompanied the overthrow of the regime further destroyed ancient records of possible astronomical worth. Despite these tragedies, some records from the earlier Shang (about 1500 to 1000 B.C.) and Chou (about 1000 to 250 B.C.) dynasties have survived.

The earliest written astronomical records are from the Shang dynasty. The people of Shang practiced divination, the supernatural prediction of the future, on a large scale. Inscriptions on tortoise shell and the bones of various animals were used. A question was written on the bones, then heat was applied to a chiseled-out cavity. The pattern of cracks resulting was then interpreted as the answer. After use the bones were buried to prevent defilement. Toward the end of last century, vast quantities of these so-called oracle bones bearing a very primitive form of ideographic writing were discovered near An-Yang in northeast China, the site of one of the main Shang capitals. They were originally sold to apothecaries as dragon bones to be ground up as remedies for various ailments. Fortunately their historical significance was soon realized. It has been suggested that the inscription on one of the oracle bones may refer to a new star that appeared near Antares; if so, this would be the earliest Oriental record of a new star. Some scholars, however, suggest that the text may be nothing more than a question as to whether a sacrifice should be made to the Great Star (Venus or Sirius?) or the Fire Star (Antares). The precise interpretation depends on a single character that could mean either *new* or *sacrifice.* Written much later, the *Spring and Autumn Annals,* possibly edited by Confucius (551–479 B.C.), contain some astronomical records, including more than thirty observations of solar eclipses.

From the Han dynasty (202 B.C.–A.D. 220) onward, we find

astronomical records of all kinds: solar and lunar eclipses, comets, meteors, planetary conjunctions (planets appearing in close proximity), occultations of stars and planets by the Moon (the Moon passing in front of stars and planets), sunspots (black spots on the face of the Sun), aurorae, sightings of Venus in daylight, certain atmospheric phenomena—and, of course, records of new stars in abundance. Later, such records were often duplicated in Korea and Japan. All in all, there are far more records relating to early astronomy extant from the Far Eastern civilizations than from any others.

The Greek and Roman classics come readily to mind as possible additional sources of historical new-star records. Again, however, we are disappointed. Certainly Pindar's "First Olympian Ode," a few lines of which were quoted at the start of this chapter, appears to refer to a new star of spectacular brightness. Even if the passage was inspired by a new star (or perhaps a comet) bright enough to be seen in daylight, poetic license has undoubtedly destroyed any astronomical value. The first known star catalog, compiled by Hipparchus about 130 B.C. and giving the positions of more than a thousand stars, was supposedly inspired by the appearance of a new star. Yet Pliny, though writing two centuries after the event, alludes to the motion of the "star," so there is a distinct likelihood that it may have been a comet. The new star that Hadrian claimed to have seen (A.D. 130) after the death of his favorite, Antinoüs, may have been only a figment of the emperor's wishful imagination. The Roman poet Claudian supposedly witnessed a temporary star that was plainly visible in the daytime and which he regarded as a portent of Honorious being made a co-emperor in A.D. 393. In this same year Chinese astronomers also recorded a spectacular new star, but unlike Claudian they left an account of its position and duration.

Two short passages in ancient Roman literature refer to celestial portents connected with the reign of the emperor Commodus (A.D. 180–192). While the Chinese did record a spectacular new star in the year A.D. 185, its position in the sky would not have been visible from Italy or Gaul, but it might have been seen from Antioch, Carthage, or Alexandria. The records merely state, "There were certain portents which coincided with these events; some stars shone continuously by day, others became elongated and seemed to hang in the middle of the sky," and "a comet appeared. Footprints of the gods were seen in the Forum departing from it. Before the war of the deserters the heavens were ablaze." It is far from obvious that either record was inspired by the A.D. 185 new star, and even if this were the case, neither contains anything of astronomical value. Indeed, it is doubtful whether there are any astronomically useful references at all to new stars in the classics. But presumably they did provide Shakespeare with the basis for his portent of Julius Caesar's death: "The heavens themselves blaze forth the death of princes."

But what of the Babylonian astronomical texts? It is well known that the late Babylonian tablets, which cover the period from about 700 to 50 B.C., contain a wealth of astronomical data on planetary motions. At this time Babylonian astrology, like its present-day counterpart, was dominated by prognostications on the basis of the positions and motions of the planets. The tablets also record lunar and solar eclipses, occultations, and a very few comets. So far, though, none of the texts studied makes reference to new stars. A vague allusion on an early tablet to a great star in Vela, supposedly originating from observations a full two millennia earlier, has been obliquely associated with the Vela supernova remnant, but this suggestion remains controversial. The lack of convincing mention of new stars is possibly a result of the fragmentary condition of the tablets, the principal collection of which is in the British Museum. The tablets represent a far from complete astronomical record for the period of interest, and in any case, bright novae as well as supernovae are rather rare. Future work on the tablets may yet reveal something of interest.

Later Middle Eastern chronicles and astrological works are also fragmentary. The Arabs' outlook was restricted by the influence of Aristotle, but medieval Arabian astronomy was far superior to its European counterpart. A number of accounts mention the 1006 new star, but surprisingly only a single record, hidden away in the journal of a physician, survives for the new star of 1054, just 48 years later.

So much for the written record of new stars. Are there unwritten sources that point to the impact of spectacular stellar events on ancient civilizations? The best example of an unwritten record of an astronomical phenomenon is the representation of the 1066 passage of Halley's comet in the Bayeux tapestry depicting the Norman invasion of England. Could spectacular supernovae have been similarly recorded? One intriguing suggestion is that the common occurrence in American Indian rock art of a crescent together with either a disk or cross may represent the new star of 1054. Modern calculations have shown that on July 5, 1054, one day after the discovery of the new star in China, the crescent Moon would have been viewed in close proximity to it from western America. Other pieces of circumstantial evidence are relevant. Every site showing the crescent/disk symbol so far investigated has a view of the eastern horizon, which would have been necessary for viewing the lunar/new star conjunction in the early morning sky. Archaeological dating at the various sites shows that they were occupied during the eleventh century.

Despite these shreds of circumstantial evidence, contrary conclusions have been suggested. Some anthropologists believe that the symbols were merely clan symbols or marked sun-watching stations—hence the view of the eastern horizon. The combination of star and crescent Moon appears frequently in ancient cultures, possibly representing occultations of bright stars or planets by the Moon and commemorated for astrological reasons. And if the American Indians were so impressed by the new star of 1054 that they depicted it in art form, why did they not leave records of the much more spectacular event just 48 years earlier? It would certainly be exciting if it could be proved that the rock art celebrated the 1054 outburst—but unfortunately doubts remain.

An interesting speculation is that the ancient aboriginal tribes of Australia may have recorded supernovae in rock art. The closest supernova to the solar system during the last 10 millennia or so is believed to have been that in Vela, at a distance of some 1,500 light-years, about a quarter the distance to the 1054 new star, and half that of the 1006 star. The Vela supernova would have passed almost overhead in Australia and would have remained visible for many years. The Aborigines are thought to have reached Australia via a land bridge from Indonesia some 30,000 years ago. Although they had no written language, decorative carvings in wood and in rock record their mythology. Included in the rock art are the so-called sunburst carvings depicting rays radiating from a central point, a familiar schematic for the sun or a bright star. Could these sunbursts be memorials to bright southern superstars, such as the nearby event in Vela?

A Roman gold coin struck by the Byzantine emperor Constantine IX, who reigned from 1042 to 1055, shows a bust of Constantine flanked by two large stars. Usually Byzantine gold coins of this period depict the emperor holding an orb in his left hand and a scepter in his right. Could one of the bright stars represent the new one of 1054, with the inclusion of the second bright star recalling the similar object of 1006? Both events could have been seen from Italy. Again, this intriguing question is unlikely to find a definite answer.

Supernovae are expected to occur close enough to the Earth to rival the quarter Moon in brightness about once every millennium. Certainly the new stars of 185 and 1006 were of such brightness. Primitive cultures surely must have been impressed by the appearance of "two moons" in the sky, with the heavens seemingly alight, and they may have felt compelled to record the event. Perhaps somewhere an as yet undiscovered engraving in stone, an uninterpreted ancient memorial, or an undetermined alignment from a neolithic astronomical site may hide a primitive message for modern science.

In discussing the written records of new stars, those of 185, 393, 1006, 1054, 1181, 1572, and 1604 have been mentioned. All these events are now believed to have been supernovae. To determine whether a new star was a nova or a supernova, we would ideally like to find in the historical record a detailed description of position, changes in brightness, and duration of visibility. Early records are rarely that complete, but whatever information about position we can obtain from the early observations helps determine whether a new star was a nova or supernova. This is because at the site of an ancient supernova explosion, one would expect to see debris from the explosion, the influence of its expanding shock wave, and possibly a stellar remnant. A postnova, on the other hand, might be identified as a close binary containing a white dwarf, possibly showing evidence of the small amount of material ejected in the outburst.

Several of the historical records of new stars refer to extreme brightness. When it comes to changes in brightness and duration of visibility, modern investigations show that supernovae and novae display characteristic and rather distinct light curves (the light curve of an astronomical object is a record of how its light intensity changes with time). In general, novae tend to show a rather more rapid decrease in brightness after reaching a maximum than do supernovae and remain above the limit of naked-eye detection for a somewhat shorter period. Unfortunately, however, it is not possible to make a distinction between novae and supernovae in our own galaxy on the basis of historical records, since the apparent luminosity of novae occurring close to the Earth would not differ from that of very much brighter supernovae at greater distance. An analogy might be sighting a source of light in the dark—could one positively distinguish between the low-power output of a nearby torch and the high-power output of a distant searchlight?

No supernova has been detected in our own galaxy since that in 1604. Therefore, present-day astronomers concentrate on the study of supernova remnants, including pulsars, expanding optical nebulosity, and extended radio and X-ray sources. The fact that few supernovae in our own galaxy have been recorded historically is partly a consequence of the dust that, as noted in Chapter 1, permeates the Galaxy and blankets many such events, rather in the way fog restricts us to seeing the headlights of only nearby cars. A large portion of the Milky Way lying toward the Galactic center, which is located in the constellation Sagittarius, is effectively screened from us optically. The situation is not improved by the Sun's location almost exactly on the plane of our flattened, discus-shaped galaxy, where the concentration of dust is highest. Only the nearby supernova remnants can be observed optically and in X rays, although the nature of obscuration is different in each case. Radio waves, however, are not obscured significantly and may be detected from supernova remnants on the remote side of the Galaxy. Out of 130 extended radio sources now believed to be the remnants of galactic supernovae, only about 30 have been detected optically and in X rays.

To appreciate the nature of the emission from supernova remnants, it is necessary to understand the electromagnetic waves that constitute light, outlined in Appendix 7. Like all waves, light is characterized by the distance between adjacent crests (the wavelength), by the amplitude of the wave, and by the number of crests passing a stationary observer per second (the frequency). The wavelength of a particular light wave determines its color. Light may be emitted from astronomical objects when electrons and ionized atoms recombine; the light thus radiated varies in color depending on the different elements making up the emitting gas. Visible light comprises just a small portion of what is called the electromagnetic spectrum. This spectrum extends from radio waves (with wavelengths of meters), through microwaves (millimeter wavelengths) and infrared radiation (submillimeter wavelengths), to visible light (wavelengths of a few one-hundred-thousandths of a centimeter), to the shorter wavelengths of the ultraviolet band, and beyond this to X rays and gamma rays. When these forms of radiation were discovered, it was not understood that they were all electromagnetic, differing only in wavelength and frequency. The unique quality of visible light is that this is the band of peak emission from the Sun that penetrates the Earth's atmosphere. Evolution has ensured that it is the radiation to which the human eye is sensitive.

In present-day astronomy, the detection and analysis of all types of electromagnetic radiation are important. Supernova remnants can produce every type, to varying degrees, and each tells us something about the nature of remnants. Of the various types of electromagnetic radiation, only visible light, some infrared and a small amount of ultraviolet, and radio waves from outer space can penetrate the Earth's atmosphere to be detected at ground level (see Figure 5). Astronomical observations at other wavelengths have been made possible over the

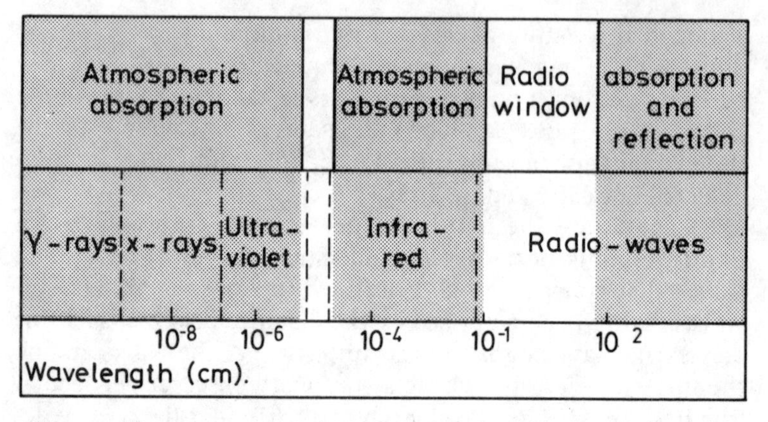

Figure 5. The electromagnetic spectrum. Visible light forms the narrow band between the infrared and ultraviolet radiations.

past two decades by lifting equipment above the atmosphere on high-flying jets, balloons, rockets, or satellites. We will now consider the radio, X ray, and optical emission from supernova remnants in turn.

Radio astronomy, which enables scientists to observe the heavens by means of radio waves emitted from astronomical objects rather than by their visible light, evolved dramatically after World War II with the development of new instruments for radio receivers and aerials. Here was technology extending the range of the human senses to permit a new look at the universe. The first discrete object (other than the Sun) identified as a radio source was the Crab Nebula, the bright nebulosity in Taurus that had been proposed as the remnant of the new star of 1054. That supernova remnants are one class of Galactic radio source was thus established almost from the beginning of the science of radio astronomy.

The discrete celestial radio sources emit a smooth continuum of radio-frequency energy; in other words, the intensity of the radio emission varies smoothly with the changing wavelength of the emission. Furthermore, some sources show enhanced emission at certain wavelengths, which is called line emission. Continuum radio emission from within our galaxy is known to come from two main types of extended object. The first of these consists of clouds of gas (mainly hydrogen) ionized by ultraviolet radiation from a bright central star; the nature of the so-called thermal radio emission from such regions is considered in Appendix 8.

The second main group of extended radio objects consists of the supernova remnants. The earliest observations of the Crab Nebula radio source showed that the intensity of the emitted energy increased with increasing wavelength. This is referred to as nonthermal radiation to distinguish it from the thermal radiation described in Appendix 8. Radio sources later identified as the remnants of Tycho's (1572) and Kepler's (1604) supernovae were also found to emit nonthermal radiation. How such emission was produced remained one of the major mysteries of radio astronomy for years. Eventually it was realized that the emission must be produced by the so-called synchrotron mechanism. The synchrotron accelerates charged particles (electrons, protons, and ions) to high energies in experiments in fundamental nuclear physics. When, in a synchrotron, electrons are accelerated to very high speed in a magnetic field, they radiate electromagnetic energy at a variety of wavelengths. The mechanism is the same in supernovae. High-energy electrons spiraling along magnetic field lines radiate light, while those of low energy radiate at radio wavelengths. The synchrotron mechanism is shown in Figure 6. The intensity of synchrotron emission characteristically decreases with decreasing wavelength. An additional characteristic of synchrotron radiation is that it displays a high degree of polarization—in other words, the direction in which the electric component of the electromagnetic field oscillates is constant. Radio emission from supernova remnants is, as would be expected, polarized. The origin of the high-speed electrons and magnetic fields in supernova remnants is considered in Appendix 9.

Supernova remnants were the only sources of celestial X rays predicted theoretically before the advent of space astronomy and the discovery of the first X-ray object (other than the Sun) in 1962. A decade and half of observational X-ray astronomy has revealed a whole range of Galactic and extragalactic X-ray

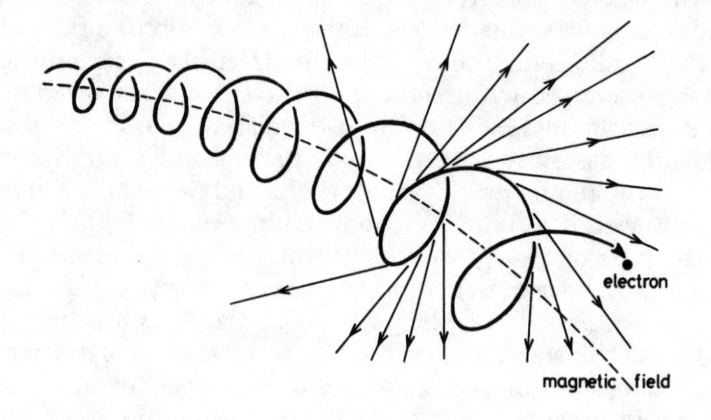

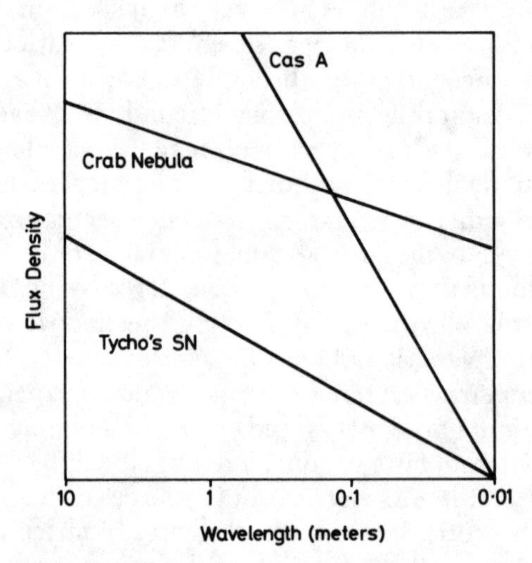

Figure 6. Synchrotron emission. (Top) Electrons in a magnetic field. (Bottom) Synchrotron emission in supernova remnants.

sources. Only a very small fraction of these sources consists of extended supernova remnants, although many other sources are believed to be systems including the stellar remnant of a supernova (a neutron star). X rays from supernova remnants are caused by thermal radiation from gas lying between the stars being swept up and heated to several million degrees by the shock wave expanding from the supernova explosion. Furthermore, synchrotron radiation is recognized as the principal source of X-ray emission in the Crab Nebula, although this X-ray-emission mechanism has been confirmed for very few other supernova remnants. X-ray sources involving a compact stellar remnant may be binary systems. Just as in a close binary system consisting of an ordinary star and a white dwarf, material is transferred to the white dwarf, so in a binary system including a neutron star or black hole, material is transferred to the compact object via an accretion disk. In the process it is heated to the extreme temperature needed to produce X rays. In a few cases, X rays have been detected directly from the hot surface of a cooling isolated neutron star at the center of a young supernova remnant.

There are three main classes of optical emission nebulae in our galaxy, each produced by essentially the same atomic processes. The word *nebula,* derived from the Latin word for cloud, is generally used in astronomy for any object that appears extended and fuzzy through a telescope. Clouds of ionized gas, made up mainly of hydrogen, represent the first class of nebulae within the Galaxy. As mentioned earlier, the ultraviolet emission from a hot central star will ionize the cloud. Recombination of an electron and an ionized atom can then take a variety of forms, resulting in the emission of light of various characteristic colors (see Appendix 10). Some well-known examples of ionized hydrogen clouds are the Great Nebula in Orion and the Lagoon and Rosette Nebulae. The second class of emission nebulae are the planetary nebulae. Spherical shells of gas believed to be emitted by stars during their red-giant phase, or alternatively by novae, may be ionized by emission from the central star. Recombination of electrons and atoms in the shell can then be detected by emission at wavelengths characteristic of the chemical elements in the shell. Examples of planetary nebulae are the Ring Nebula in Lyra, and the Helix and Dumbbell Nebulae. The third main class of emission nebulae are the optical remnants of supernovae, where the source of energy is not necessarily a central star. In the case of the Crab Nebula, ionization is produced by synchrotron radiation from the central pulsar. For the majority of optical remnants, however, the ionization is believed to be produced by the expanding shock wave. We have noted that the interstellar medium is not uniform but is believed to be clumpy, containing clouds of enhanced density. As the shock blasts its way through this clumpy medium, the clouds are heated and ionized, then as they cool, through about 10,000 degrees, free electrons can be recaptured by the ionized atoms, with the emission of visible light. Nature's brush has painted the heavens with a host of beautiful nebulae, using a palette of many colors, which characterize the elements. Among the most spectacular are those associated with the supernovae.

So much for the extended remnants of supernovae—but what about their stellar remnants? The exact reason why pulsars emit radiation is still uncertain. When a star collapses, its magnetic field is greatly compressed. Just as a pirouetting ice skater spins faster as he brings in his arms, so a slowly rotating star would spin more as it collapsed to a neutron star. Thus a recently formed neutron star (with rotation period of just a fraction of a second, compared with the many-day rotation period of the presupernova star) resembles a rapidly spinning magnet that radiates electromagnetic waves. Somehow—and theoreticians admit they do not yet know quite how—this radiation is concentrated in a narrow, coherent beam, which flashes around the sky like the beam of a lighthouse. If the beam sweeps over the Earth, which is believed to happen for about one in four pulsars, the neutron star will be observed as a pulsar (see Figure 7) with a pulse period indicating the rotation period of the neutron star.

Most of the cataloged Galactic supernova remnants are probably so old that no records of the explosions that created them can be expected to exist, even for those that occurred close

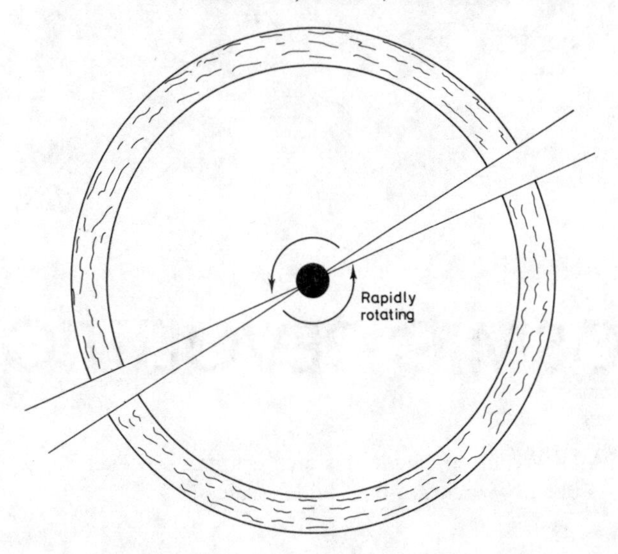

Figure 7. Beamed radio emission from a rapidly rotating neutron star at the center of an expanding supernova remnant.

enough to the Earth to have been witnessed. For this reason, the few available historical observations of supernovae, crude as they are by present-day standards, are very important. If a particular remnant can be associated with certainty with a new star recorded in a particular year, then the period during which the remnant has been developing is precisely known. Several such identifications could provide valuable observational evidence to test current theories on the evolution of supernova remnants and on the nature of supernova explosions. There can be little doubt that the historical new-star records must be regarded as among the most valuable legacies of the ancient world. But it is not merely the association of historical records of superstars with presently observable remnants that make superstars important. At the time of their outburst, superstars contributed to advances in cosmology, helping to fire the revolution that overthrew the old view of the universe and providing the foundation on which to build a new view.

3

COSMIC REVOLUTION

All destruction, by violent revolution or however it be, is but new creation on a wider scale.

—*Thomas Carlyle*

THE IDEAS OF TWO MEN, Aristotle and Ptolemy, dominated classical astronomy. The concept of a perfect universe was formulated by Aristotle. Subsequently, in the second century A.D., Ptolemy of Alexandria became one of the best-known proponents of the so-called geocentric theory that the Earth lay at the center of the universe while the heavens moved around it fixed to a system of invisible, crystal concentric spheres rotating at different speeds. The Moon occupied the innermost sphere; then came the planets Mercury and Venus; then the Sun, followed by the planets Mars, Jupiter, and Saturn. The eighth and outermost sphere contained the multitude of fixed stars. To the ancient Greeks, a sphere represented spatial perfection, so that a perfect-universe model based on spheres seemed obvious. The invisible, crystal spheres were needed to stop the stars and planets from falling to Earth! In fact, no single set of spheres could possibly describe the complex wanderings of the planets amongst the background stars. Attempts to overcome such difficulties involved attaching secondary rotating spheres at varying angles to the primary spheres of the planets.

Despite these problems Ptolemy clothed the theory in a complex mathematical formulation; the observational data he used was from a variety of sources, among them the original star catalog of Hipparchus. The result of Ptolemy's labors was his monumental work *The Great Synthesis,* commonly called *Almagest* from its Arabic translation. Modern calculations suggest that Ptolemy may have fabricated some of the observational results to fit them to his theory—he could never have dreamed that history, armed with telescopic observations of extreme precision and calculations made with electronic computers, would make such a harsh judgment on his masterpiece. But regardless of whether Ptolemy "fiddled the books" or not, the impact of *Almagest* on medieval thought was truly astounding. Few modern theories based on observations of extreme precision involving advanced instrumentation and sophisticated computation could hope to match its protracted influence. The geocentric model of the universe remained essentially intact through the dark ages of scientific ignorance following the collapse of the Greco-Roman culture. By the twelfth century it formed a cornerstone of the religious dogma of the Church of Rome, so that the questioning of the Church's interpretation of either the Aristotelian doctrine (a perfect, changeless, celestial vault reflecting a perfect deity) or the Ptolemaic system (an Earth created by God especially for humankind, at the center of the universe) became tantamount to heresy.

The first serious alternative to the Ptolemaic theory appeared when the Polish cleric Nicolaus Copernicus published his theory of a Sun-centered system in the monumental treatise *De revolutionibus orbium coelestium* ("On the Revolutions of the Heavenly Bodies") shortly before he died in 1543. What Copernicus had proposed was to rock the foundations of Western scientific thought and test the infallibility of papal doctrine in the century to follow. The very word *revolution,* when applied to the violent overthrow of an existing system, has its origins in the treatise.

Copernicus initiated the great revolt in scientific and religious beliefs in the sixteenth and seventeenth centuries by placing the Sun at the center of his system and arranging the six planets then known (Mercury, Venus, Earth, Mars, Jupiter, and Saturn) in circular orbits around the Sun. We now know that his circular orbits were a simplified representation of reality and that he made several other errors of interpretation, but the Sun-centered system was simpler than the Earth-centered one and correctly described the general planetary motions. A more exact interpretation, involving elliptical orbits for the planets, emerged from the work of Johannes Kepler some seventy years later. The problem was that the Copernican demotion of the Earth (and consequently its inhabitants) from its central, "all-powerful" position in the universe was seen as an affront to a theology and social order evolved over 1,500 years. Although the Church and conservative intelligentsia tried hard initially to ignore Copernicus's work, its growing influence on free thought eventually resulted in it being placed in 1616 on "The Index of Prohibited Books," where it remained for over two centuries.

If Copernicus had produced the first cracks in the crystal spheres, they were finally shattered by the appearance of the new stars of 1572 and 1604. The stars could not have appeared at a more critical time in the development of astronomical thought in Europe. One of the basic tenets of the Aristotelian perfect universe was that the eighth sphere of the fixed stars was immutable. The earlier appearances of new stars recorded in the West were dismissed as being sublunary—in other words, lying below the first of the crystal spheres occupied by the Moon and within the zone subject to change. It was therefore critical to have accurate positional measurements of the new star of 1572 in order to determine whether it could possibly belong to the eighth sphere. The best-known positional estimates were those made by Tycho Brahe. In a preliminary report entitled *De nova stella,* published in 1573, Tycho discussed his measurements of the position of the star in relation to the principal stars in the constellation Cassiopeia, within which the new star appeared. He concluded:

> That it is neither in the orbit of Saturn, however, nor in that of Jupiter, nor in that of Mars, nor in that of any one of the

> other planets, is hence evident, since after the lapse of several months it has not advanced by its own motion a single minute from that place in which I first saw it; which it must have done if it were in some planetary orbit. . . . Hence this new star is located neither in the region of the Elements, below the Moon, nor in the orbits of the seven wandering stars, but in the eighth sphere, among the other fixed stars.

It is possible that even in the newly enlightened academic climate of the sixteenth and seventeenth centuries a single new star among the fixed stars might not have destroyed the idea of the perfect universe. Just one such object might have been accommodated within the protestations of the Aristotelians as being a special case, miraculous (and heralding the Second Coming?) or merely an illusion. Alternatively, the positional measurements that placed the new star in the eighth sphere might have been dismissed as being in error. The appearance of another spectacular new star shortly after offered an opportunity to test this last possibility and dismiss the others.

Many people who witnessed the 1572 event must also have seen the new star which occurred just thirty-two years later. Unfortunately, Tycho Brahe did not, for he had died in 1601. Johannes Kepler, however, who was his successor as Imperial Mathematician at Prague, was to make similar detailed observations of the new star, which bears his name. Kepler is best known for using the results of observations of Mars made by his old master, Tycho, to establish that the planets moved in elliptical orbits, with the Sun at one focus, and the speed of a planet varying around its orbit in a well-determined way. But of interest to us here are his accurate positional determinations of the new star of 1604 relative to the other stars in the constellation Ophiuchus, in which the new star appeared. This information he published in his *De stella nova in pede Serpentarii.* Positional estimates of similar precision were made by David Fabricius and others. Again there could be no doubt—the new star lay beyond the orbits of the planets, among the fixed stars. The case was now all but proven. The universe was not perfect but was able to bring forth "new" stars as spectacular manifestations of its ability to change. The new stars had helped shatter not only the crystal spheres but also a social order established over nearly two millennia. People were forced to look anew at their science, religion, philosophy, and cosmology.

Almost three hundred years passed before new stars again featured prominently in forming people's ideas of the cosmos. In the intervening period, the advent of the telescope had confirmed the Copernican concept, modified by Kepler, of a heliocentric planetary system and had revealed such phenomena as imperfections on the Sun (sunspots) and variable stars (stars that brighten and fade in a regular manner) as further armament to be directed against the Aristotelian dogma of heavenly immutability.

The distinction of inventing the telescope probably lies with a Dutch lens maker. It was Galileo Galilei, however, who improved the magnification to the point where it could be used as an instrument for astronomical observation. Galileo first turned his telescope to the heavens in 1609, marking the birth of modern observational astronomy. Shortly after, he wrote in his book *The Starry Messenger:* "I have seen stars in myriads which have never been seen before, and which surpass the old, previously known stars in number more than ten times." Galileo's observations confirmed that the Milky Way, the nebulous band stretching across the heavens, was made up of a multitude of closely packed, faint stars, an idea suggested as early as 425 B.C. by the Greek philosopher Democritus. Galileo became an outspoken proponent of the Copernican system, despite warnings by the Church. In his famous dialogue *The Two Chief World Systems,* published in 1632, he not only questioned Aristotelian and Ptolemaic doctrines but also appeared to ridicule established Church authority. The book was banned, and he was brought to trial in Rome in 1633. He was forced, under threat of torture, to acknowledge publicly that his ideas were wrong. However, his humiliation could not stop the eventual acceptance of his revelations of the true nature of the heavens. Galileo, now blind, lived out his remaining few years under house arrest, a broken and bitter man. The Church's stance stifled much free scientific thought in southern Europe, and major post-Renaissance advances in science were to gel in northern Europe. Developments on Galileo's hypothesis that the Milky Way was made up of a multitude of faint stars were in fact made in England.

William Herschel is acknowledged as one of England's greatest observational astronomers. He was a German by birth (in 1738), but immigrated to England in 1757 from Hanover, where he had been an oboist in the Hanover Guards. He remained a professional musician in his newly adopted land but pursued astronomy as a spare-time activity. While doing so in 1781, he accidentally discovered a new planet, later named Uranus. This single discovery gained him the royal patronage of King George III, enabling him subsequently to spend all his time on astronomy. One of Herschel's principal interests was to determine the structure of the Milky Way. This involved him, with the help of his sister Caroline, an astronomer in her own right, in a lengthy program of star gauging—counting the number of stars in certain selected regions. By assuming that the observed brightness of a star depended only on its distance—an assumption we now know to be invalid—and ignoring absorption of starlight, Herschel asserted that those regions containing the greatest number of faint stars corresponded to the maximum extensions of the system of stars. From such arguments Herschel concluded that the Sun was near the center of a gigantic, flattened conglomerate of stars, the Galaxy which is shaped like a biconvex lens but with irregular boundary. (*Galaxy* comes from the Greek word for Milky Way, *galaxias.*) The star-gauging technique was refined in the latter half of the nineteenth century by the Dutch astronomer Jacobus Kapteyn, who allowed for stars at similar distances having differing intrinsic brightness based on studies of stars in the solar neighborhood. Although the Kapteyn universe had a smoother periphery than that determined by Herschel, it retained the disk structure for the Galaxy with the solar system close to its center.

Apart from the Sun's position, Herschel's model of the Galaxy was to prove later to be very nearly correct, as was another of his assertions. A catalog of a hundred small, faint clouds of emission, called nebulae, which might cause confusion in cometary studies, had been prepared by the famous comet-hunter Messier. The Herschels' star gauging had revealed some 1500 more such objects, many of which had a characteristic structure—the spiral nebulae. Because most of these seemed to lie away from the plane of the Galaxy, and some could be resolved into individual stars, William Herschel suggested that all the nebulae might be separate, distinct galaxies lying far beyond the Milky Way. This brilliantly intuitive guess made little impression at the time, and even up until the early twentieth century many scientists believed that our galaxy represented the entire universe. (The concept of a universe made up of many galaxies had been alluded to in the mid-eighteenth century by the English scientist Thomas Wright and German philosopher Immanuel Kant, who termed the nebulae *island universes*). Herschel himself helped to undermine the nebula/galaxy hypothesis with his work on certain nebulae that appeared to be gaseous disks (planetary nebulae) rather than star conglomerates. An alternative interpretation of the bulk of the nebulae, as solar systems in the making, gained popularity throughout the nineteenth century.

The interpretation of nebulae lying within the Milky Way and being solar systems in formation was consistent with a theory of the origin of our own solar system developed by Pierre Simon de Laplace toward the end of the eighteenth century. In this theory, which forms the basis for currently held views on the formation of the solar system, the Sun contracted from an immense rotating cloud of gas. During the contraction the rotation sped up, so that matter spread in a disklike fashion, and the central cloud eventually contracted to form the Sun and the disk broke up to form the planets. The nebulae were similarly believed by some to lie within the Milky Way and to be the final stages in the evolution of immense clouds into solar-type systems. Did not many nebulae—for example, the famous Andromeda Nebula (just visible to the naked eye and recorded since antiquity)—look like giant, rotating, spiral disks? Early photographs showed that the Andromeda Nebula had two detached satellite objects—could the nebula be a nascent solar system within the Milky Way and the satellites be planets in the making?

A spectacular new star near the center of the Andromeda Nebula in 1885 seemed to be accommodated by the Laplace theory. Credit for the discovery of the 1885 new star has been given to the German astronomer Ernst Hartwig working at the Dorpat Observatory in Russia. He was certainly not the first to see it, but he was definitely the first to realize its significance, believing first of all that it was a central Sun appearing within the nebula in accordance with the Laplace theory. Its subsequent decline in brightness, so that it could no longer be distinguished after a period of 180 days, disproved Hartwig's idea. Nevertheless, the 1885 new star and another discovered ten years later in a nebula called NGC 5253 (object number 5253 in the *New General Catalogue* of nebulae) were used in renewed attempts to decide whether the spiral nebulae were external galaxies or local star groups within our own galaxy. As it turned out, their contribution to this debate was to prove highly misleading.

Yet another new star was to play a part in unraveling the mystery of the nebulae. In 1901 a nova appeared in the constellation Perseus, brightening to become one of the most spectacular objects in the heavens within just a few days of discovery, before declining into insignificance over a period of merely a few weeks. Shortly after its decline, the site of the nova was found to be surrounded by a small, faint nebula. It was impossible in the time since the nova outburst for ejected material to have reached the distances implied by the appearance of the nebula; a more likely explanation was that the burst of light from the nova illuminated successive layers of surrounding dust, producing a so-called reflection nebula. The reflection nebula was therefore expanding with the velocity of light. As it expanded, it gradually faded. By comparing the rate of its increasing angular size on the sky with its known expansion velocity (the speed of light), it was possible to estimate its distance. This was found to be about 5 million billion kilometers. In other words, Nova Persei was estimated to be at a distance of about 500 light-years.

The apparent brightness of a symmetrically radiating light source diminishes as the inverse of the square of its distance from an observer; the effect is referred to as the inverse square law. Nova Persei had been about 250 times brighter at its maximum light than the new star of 1885 in Andromeda. If, then, both new stars were similar in type, the Andromeda Nebula must be a mere 16 times more distant than Nova Persei—that is, at 8,000 light-years from Earth, and well within the confines of the Milky Way as its size was then understood. Unfortunately, however, this was not the end of the nebula mystery, and a solution was still a full two decades away. Several other new stars were discovered in the Andromeda Nebula, but none of comparable brightness to that of 1885. Heber D. Curtis insisted that the new star of 1885 had therefore been unusual, and suggested that two types of nova existed—normal (such as Nova Persei) and superbright (such as that of 1885). Curtis insisted that if the 1885 new star was ignored, the other new stars in Andromeda implied a distance consistent with an island-universe classification. Novae discovered in other spiral nebulae appeared to force them also to distances far beyond the extremities of the Milky Way. The Swedish astronomer Kurt Lundmark independently proposed that rare events some 10,000 times brighter than ordinary novae also occur.

As an important and continuing part of the nebula debate, parallel investigations were being carried out into the size of the Milky Way. Although by the turn of the twentieth century it was believed that the Sun was just one of the vast multitude of stars making up our galaxy, there was still no consensus on the true size of the Galaxy. Whether the universe extended beyond the Milky Way was open to speculation, but the Sun was thought to lie somewhere near its center. Soon, however, the Sun was dislodged from this preeminent position, when Harlow Shapley subjected it to the same humiliation as the Earth had suffered at the hands of Nicolaus Copernicus some 350 years earlier. Shapley's early career hardly suggested that he would initiate a revolution in our understanding of the universe. He started as a crime reporter for a small-town newspaper in Kansas, covering the fights of drunken oilmen. Wishing to better himself, he entered the University of Missouri, hoping to study journalism. However the school of journalism had not yet opened, and an alternative field of study had to be selected; as Shapley describes it in *Through Rugged Ways to the Stars:* "I opened the catalogue of courses. . . . The very first course offered was a-r-c-h-a-e-o-l-o-g-y, and I couldn't pronounce it! . . . I turned over a page and saw a-s-t-r-o-n-o-m-y. I could pronounce that—and here I am." Shapley was a determined, self-confident man who enjoyed an argument. A less strong personality might not have been able to force radical new ideas upon an initially skeptical scientific community.

The banishment of the Sun to the outer reaches of our galaxy had its beginnings in the sixteenth- and seventeenth-century discoveries of variable stars. Then in 1912 Henrietta Leavitt of the Harvard Observatory discovered that variable stars in the Magellanic Clouds (two satellite galaxies to the Milky Way) showed a relationship between their period (the interval between times of peak brightness) and their average apparent brightness. The particular subgroup of variable stars studied had a period of about 1 to 30 days and are called cepheids (see Figure 8). Leavitt's examination of a period-brightness relation suggested that, since all the Magellanic Cloud variables were at nearly the same distance from us, the real brightness of a cepheid is linked with its period of variation. A cepheid with a

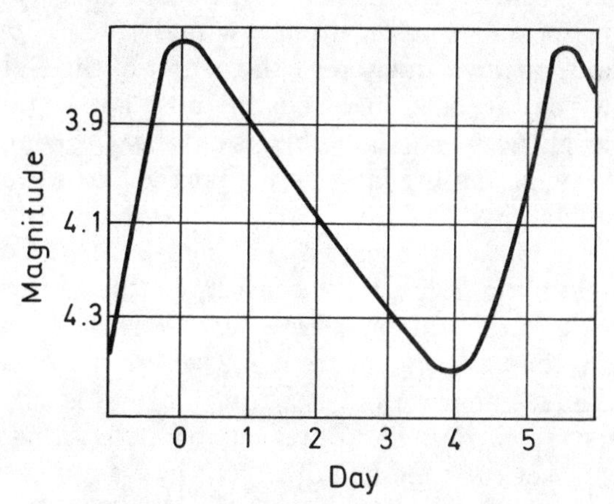

Figure 8. The regular fluctuations of the light curve of the variable star Delta Cephei.

period of 10 days will have a greater intrinsic brightness than a variable with a period of 5 days, and so on. The intrinsic-brightness scale was then calibrated by Shapley from observations of nearby cepheids in our own galaxy for which distances could be determined. The period-brightness relation therefore made it possible to estimate the distance to any cepheid with a well-determined period and average apparent brightness, by application of the inverse square law.

It was noted that a certain type of variable star with a short period of about 1 to 30 hours was often associated with globular clusters—dense groups of many hundreds of thousands of stars—within the Galaxy. Cluster variables are usually called RR Lyrae variables. Shapley extrapolated the use of the period-brightness relation to RR Lyrae variables to determine their distances. The globular clusters were found to be scattered throughout a flattened spherical volume with a diameter of about 100,000 light-years, with the Sun about 30,000 light-years from the center of the distribution, which was presumably the center of the Galaxy. Since 30,000 light-years is about 300 million billion kilometers, Shapley's calculations had revealed the mind-boggling extent of the Galaxy and forced the Sun closer to its periphery than to its heart.

What had been wrong with Herschel's and Kapteyn's deductions from the star-gauging studies, which gave the erroneous impression that the Sun lay near the center of the Galaxy? We now know that absorption by cosmic dust between the stars dims the brightness of distant objects close to the central plane of the Galaxy, giving the false view that the Earth lies near its center (see Figure 9).

In 1920 the National Academy of Sciences arranged a meeting in Washington, D.C., to discuss the nature of the nebulae and the scale of the universe. How large was the Galaxy? Did the universe extend beyond it? If so, were the spiral nebulae island universes or were they part of our own galaxy? Curtis and Shapley were invited to present the two points of view. From the outset the men had divergent views not only on the scientific issues but also it seems on what the subject of the

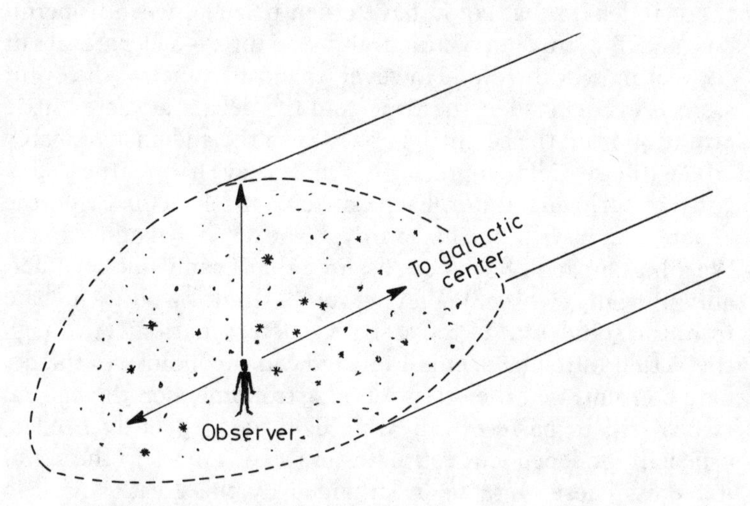

Figure 9. Interstellar absorption in the plane of the Milky Way.

debate was supposed to be. Shapley wanted to limit the debate to the size of our own galaxy, but Curtis insisted that the nature of the spirals be included. When the debate took place, Curtis's presentation was dominated by the nature of the spirals and his insistence that they were distant island universes—while Shapley spent just a few minutes arguing that the spirals were much closer than Curtis proposed and much smaller than the Galaxy. Shapley concentrated his efforts in arguing for a larger Galaxy than had hitherto been accepted, with a structure differing dramatically from the then-favored Kapteyn view. The difference in emphasis of the two debators was to be expected. Shapley had spent many years investigating the size and structure of the Galaxy, while Curtis had spent as long photographing and studying the spirals. The level of presentation also differed. Curtis discussed highly technical evidence, while Shapley, better sensing the varying interests of an audience coming from all areas of science, adopted a more general approach. In historical perspective, the encounter seems important, and it has been compared by some to a hypothetical debate that could have taken place between Copernicus and Ptolemy, or Galileo and Aristotle—a debate about competing world views. However, the importance of the event went unrecognized at the time, and the debate attracted little attention from the scientific press. Even the audience probably left without fully appreciating that they had witnessed a truly historic encounter. Each participant felt he had won the debate. Curtis wrote to his family, "The debate went off fine in Washington, and I have been assured that I came out considerably in front," while Shapley wrote, "I think I won the debate from the standpoint of the assigned subject matter." Although the actual outcome seems to have been inconclusive, the debate concentrated the attention of astronomers on the central issues still to be resolved. More data were urgently needed, especially independent estimates of the distances to the spiral nebulae. These were to be provided by the great American astronomer Edwin P. Hubble.

Hubble, like Shapley, arrived at a career in astronomy by a roundabout route. After earning a degree in science, he was awarded a Rhodes Scholarship to Oxford, where he studied law. On returning to the United States in 1913, he spent a short time practicing law. But his first love was science, and he obtained a Ph.D. in astronomy in 1917, followed by a year of war service. He then moved to the Mount Wilson Observatory in California. In 1923 Hubble was able to use the new, giant, 100-inch telescope on Mount Wilson to take a series of high-resolution photographs of several of the spiral nebulae, most notably Andromeda. These enabled him to identify several cepheids. The period-brightness law for cepheids, together with a later recalibration of the distance scale, allowed him to place these and their parent system the Andromeda Nebula at an unbelievable distance of two million light-years—far beyond the outer boundary of our galaxy. Surprisingly , Hubble seemed at first reluctant to publish these amazing results but was eventually persuaded in late 1924 that they should be presented to a meeting of the American Association for the Advancement of Science in Washington. One of the participants of that meeting in December 1924, which Hubble himself did not attend, later recalled, "The entire Society knew that the debate (Curtis vs. Shapley) had come to an end, that the island-universe concept of the distribution of matter in space had been proved, and that an era of enlightenment in cosmology had begun." The majority of nebulae had suddenly become galaxies in their own right, some 170 years after Kant had first suggested it, and the most amazing step so far had been made in humankind's appreciation of the enormous size of the universe. Planet Earth, once residing in splendor at the center of an Aristotelian perfect universe, now found itself merely a part of a very ordinary planetary system, on the outskirts of a rather typical galaxy, which in turn was one of an apparent vast multitude of stellar systems.

Hubble's photographs of a large number of extragalactic nebulae (from now on to be referred to simply as galaxies, with the term *nebulae* to be restricted for gas clouds within a galaxy) showed that no two were identical in form. Nevertheless, he was able to confirm the existence of certain distinct classes inferred, with less certainty, from earlier studies. Elliptical galaxies have an oval shape; spiral galaxies have a bright central region, or nucleus, often appearing as a bright "bar," with two or more trailing spiral arms; finally there are the irregular galaxies, which cannot be classified in any of these ways. As noted earlier, the galaxies tend to congregate in groups, or clusters. The Milky Way is part of the so-called local group of about 20 galaxies (including the Andromeda Nebula), although some huge clusters of galaxies may have a membership of 10,000. The average separation between galaxies is three million light-years, although within individual groups separations may be less than a million light-years.

Not all galaxies are quiescent structures. In 1943 Carl Seyfert discovered several spiral galaxies with hot gas being ejected from their nucleii at speeds of up to 1,500 kilometers per second, and it is now known that about one percent of spirals are Seyfert galaxies. It has been suggested that the violent outbursts of material from Seyfert galaxies could be caused by a series of closely spaced supernova explosions at their centers, although there are difficulties in such an explanation. Other classes of superenergetic objects are known, the most enigmatic type being the quasars, which appear to be emitting about a hundred times the light of a typical large galaxy, though this energy is concentrated in a mere fraction of the size of a typical galaxy. It is unlikely that the enormous energy of a quasar could originate from a succession of stellar explosions. Instead, it is now proposed that the energy may come from accretion of matter (stars, plus interstellar gas and dust) onto a massive black hole at its center.

The new appreciation of the true vastness of the cosmos with the majority of the "nebulae" being galaxies at extreme distances, proved conclusively that the new star of 1885 in Andromeda, like that of 1895 in NGC 5253, must have been a remarkably brilliant object, intrinsically perhaps several thousand times brighter than the ordinary novae also identified in the galaxies. Other such superluminous new stars were soon identified in other galaxies, although it was quickly realized that they were comparatively rare events. There now seemed no escaping the fact that, as Curtis and Lundmark had suggested earlier, there were two distinct subclasses of new stars. In 1937 the term *supernova* was proposed by the Swiss astronomer Fritz Zwicky and the German-American astronomer Walter Baade, two of the pioneers of modern supernova research, for the distinct group of new stars of extreme intrinsic brightness. The superstars had finally been revealed in their true glory. Along the way, the struggle to determine the structure of the Milky Way and the nature of the nebulae, begun in earnest by William Herschel, had reached a satisfactory conclusion. In this, as in the shattering of the crystal spheres, the superstars had played their part. But their role was not complete. They still had a bit part to play in building the new "expanding universe" from the shattered debris of the old.

To understand the new cosmology—the new theory of the universe—we need to digress to consider how light can be broken up into its component colors to form a spectrum. This topic is considered in Appendix 11. The different elements in their gaseous form emit light of characteristic colors, which are unique signatures of each element. Viewed directly, with the gas radiating, these colors appear as bright individual lines when the light is broken up into a spectrum (see Appendix 11); this is an emission-line spectrum. Alternatively, if illuminated from behind by a white light source, the characteristic signature of an element appears as dark lines on the continuous spectrum of the white light source; this is an absorption-line spectrum.

The spectra of stars show absorption lines, which indicate the composition of the stellar atmosphere, against a continuum background originating from the star's visible surface—its photosphere. For a conglomerate of closely bunched stars, whether a cluster or a galaxy, the main absorption lines should still be identifiable, in addition to emission lines originating in interstellar gas. When, however, spectroscopic studies of 40 spiral nebulae (subsequently realized to be nearby galaxies) were carried out by Vesto Slipher at the Lowell Observatory in Arizona between 1912 and 1920, it was found that although the principal spectral lines (absorption and emission) of these galaxies could be identified, they were almost always displaced in wavelength toward the longer red end of the spectrum. In other words, the light from these galaxies is red-shifted (see Figure 10).

A color shift of light is explained in terms of the Doppler effect (see Appendix 12). If a light source is moving away from an observer, its color is red-shifted; if it is moving toward the observer, its color is blue-shifted. The effect is only discernible at extreme speeds. The degree of blue shift or red shift provides a way to estimate the speed at which the light source approaches or recedes.

Slipher's red shifts were interpreted as motions of the galaxies away from the Milky Way. This was already strange enough, but Hubble was to make an even more startling revelation. This depended on his estimates of the distances to remote galaxies by using the brightness of their brightest stars. The dimmer the bright stars, the farther away their parent galaxy in accordance with the inverse square relation. The calculation of the distance to one galaxy using cepheid variables will then enable the distance to any other galaxy of similar type to be

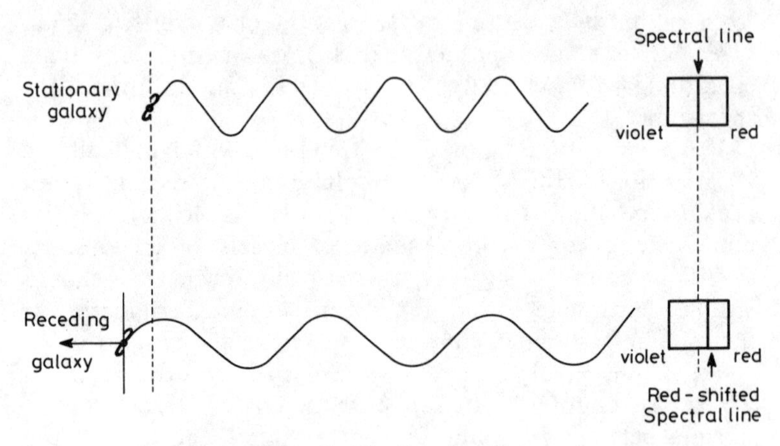

Figure 10. The Doppler effect.

determined. If, for example, the Andromeda Nebula is taken as the standard calibrator, with a distance of two million light-years estimated from its cepheids, a spiral galaxy whose brightest stars were four times dimmer than those in Andromeda would be at twice the distance, and so on. The observations were very difficult. By 1929, although the red shifts of 46 galaxies had been found, distances had been determined for only 18. Nevertheless, Hubble felt confident enough to present his results to a meeting of the National Academy of Sciences; his work seemed to demonstrate a clear relationship between the recessional velocities of spiral nebulae and their distances (see Figure 11). This relationship is now known as Hubble's law: The more distant a galaxy, the greater its red shift. Or in the law's simplest interpretation, assigning the red shift entirely to recessional speed: The more distant a galaxy from the Milky Way, the greater the speed of recession. In mathematical terms, Hubble's law may be summarized by the equation $v = Hd$, where v is the recessional speed of a galaxy at a distance d. H is called the Hubble constant, and has the approximate value

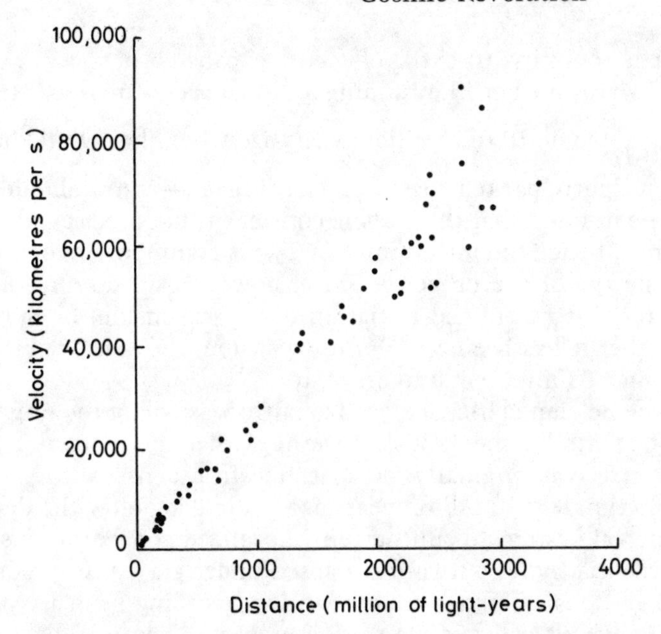

Figure 11. Hubble's relationship between recessional velocities for galaxies and their distances.

of 20 kilometers per second per million light-years. By 1936, Hubble had shown that his law applied to galaxies with distances up to 500 million light-years. Today, we believe it holds to more than 10 times this distance.

The recessional speeds of the galaxies are enormous. For example, a galaxy at a distance of 50 million light-years is moving away from us at about 1,000 kilometers per second; at 10 times this distance the speed will be 10,000 kilometers per second, and so on. Although all other galaxies are flying away from the Milky Way at great speed, this still does not mean that our galaxy is at the center of the universe, for the simplest way to interpret Hubble's law is as a general uniform expansion of the universe. Lying within this general expansion, all the galaxies appear to be receding from us; but if we were observing the universe from any other galaxy, we would witness exactly the same effect.

It is tempting to extrapolate the expansion of the galaxies backward—rather like running a movie backward—to a time $T_0 = \frac{1}{H}$, some 10 to 20 billion years ago when they would have been tightly packed together. The time T_0—and it should be borne in mind that the Hubble constant is not precisely determined, hence the uncertainty in T_0—is commonly referred to as the age of the universe. Since, however, we do not know what went before the initiation of the expansion, or indeed whether it has been a uniform expansion, the true age of the universe is impossible to estimate.

The Belgian cleric Georges Lemaître was one of the original proponents (in the 1930s) of the idea that the matter of the universe was originally concentrated in a dense form. Then approximately 10 billion years ago, a Big Bang blew the dense primeval system to smithereens. Localized concentrations in the debris flying outward collapsed under gravity to produce the galaxies of stars we now witness receding from us. The most distant galaxies we can observe with the world's most powerful telescopes would then appear to us as they were shortly after their birth; quasars are thought by some to be galaxies at this early epoch. The nearby galaxies appear as of similar age (just a few million years older) to our own. The history of the universe can therefore be revealed by the study of galaxies at varying distances (see Figure 12), as we witness the expansion of the cosmos.

Will the expansion of the universe continue forever, or will gravitational attraction between the clusters of galaxies eventually halt the expansion and indeed reverse it? The future behavior of the universe depends on the density of the material within it. As already noted, only a small fraction of the material within the universe is trapped within the stars. The rest exists as tenuous gas, lying, often in clumpy clouds, between the stars and between the galaxies. Much of this intergalactic and interstellar material is not visible to us, and it is the exact amount of this "missing mass" that will determine whether the expansion of the universe will eventually halt. If 99 percent of the mass of the universe is invisible, then eventually space

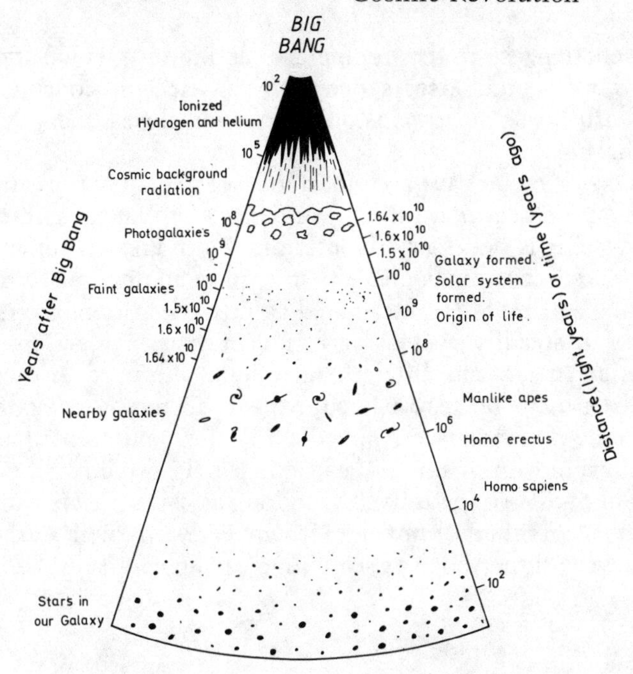

Figure 12. Schematic representation of the history of the universe.

will turn in on itself and the galaxies will be dragged toward each other in a universal contraction. Scientists speculate that such infall could eventually precipitate another Big Bang as the galaxies are crushed together, giving birth to a new expanding universe out of the debris of the old. This is the basis of the oscillating-universe theory, in which a Big Bang and an expansion phase of the universe (lasting, it is conjectured, about 100 billion years) is followed by a contraction phase of equal duration and the initiation of another Big Bang and a next-generation universe. This theory has the attraction of providing the universe with a permanency such that it requires no true beginning and no ultimate end, but comprises merely a series of creative epochs punctuated by Big Bangs and interspersed with periods of expansion and contraction—a universe with an infinite future and an infinite past. One of the most pressing challenges to contemporary astronomy is to decide whether the universe is open, so that it will continue its presently observed expansion forever, or closed and possibly oscillating.

The superstars have an important role to play in these investigations. To estimate the mean density of matter within the universe, it is necessary to determine with great accuracy the distances to cosmic bodies—that is, to accurately measure the Hubble constant. The estimated value of the Hubble constant has been steadily decreasing in recent years as new observations are made and different techniques are tried. A method has been developed to use supernovae in remote galaxies to calculate distance (see Appendix 13). It would seem that supernovae are in a unique position to help establish the true distance scale of the universe, so long as astronomers can discover a sufficient number of them early in their outburst. Hence the importance of searching for superstars.

4

SEARCHING FOR SUPERSTARS

There awaited us an unknown buried multitude of hidden treasures . . . new cosmic bodies and phenomena which could only be divined through systematically directed intuition and subsequent tenacious search with proper instruments.

—*Fritz Zwicky*

MODERN SUPERSTAR SURVEYS originated in the 1930s among a mere handful of dedicated individuals. While the controversy over the nature of the nebulae revealed the occasional appearance of new stars of exceptional brilliance, there was no clear idea of what initiated these outbursts or how often they might occur. Several were discovered in the early decades of the twentieth century, in nebulae (subsequently identified as external galaxies), but these were merely chance discoveries.

Systematic surveys of the heavens for new cosmic bodies and phenomena became possible with the building of large and new types of telescopes. In 1928, George Ellery Hale, then director of the Mount Wilson Observatory in California, wrote an article for *Harper's Magazine* entitled "The Possibilities of

Large Telescopes." His article began: "Like buried treasures, the outposts of the universe have beckoned to the adventurous since immemorial times." Rumor has it that multimillionaire industrialist and philanthropist J. D. Rockefeller read only the first few sentences of the article before telephoning Hale with an offer of $6 million through the Rockefeller Foundation. This inspired patronage resulted in the construction of the giant, 200-inch reflector telescope at Mount Palomar, which remained the world's largest telescope until recent years. A smaller (18-inch) but powerful survey camera, known as a Schmidt telescope, was built at the same time to act as a scout instrument for the 200-inch telescope. Schmidt telescopes are constructed to produce excellent images over a very wide field of view, usually several degrees across, compared with the usable field of view of the 200-inch of just a small fraction of a degree across. They can therefore survey a larger portion of the sky at one time than conventional telescopes. In the mid-1930s, Fritz Zwicky and his collaborators were to prove the Schmidt telescope one of the most powerful innovations in observational astronomy.

Zwicky was one of astronomy's more controversial personalities. Dedication to the cause of "directed intuition"—the so-called many-sided or morphological method of research—resulted in his making predictions about astronomical phenomena, and supernovae in particular, which most of his contemporaries found to be totally unacceptable. Zwicky, in collaboration with Walter Baade, argued as follows:

1. Because of attractive gravitational forces, matter in general has a tendency to agglomerate and to compact.
2. The tendency toward compaction is counteracted by the energy possessed by the various bodies in the universe.
3. When it occurs, compaction (or implosion) cannot proceed unilaterally, since in the process energy will be released which automatically results in some matter being ejected at high speeds (an explosion).

Zwicky and Baade believed that the superluminous novae in external galaxies, the supernovae, were caused by the collapse (implosion) of stars or clouds of gas and dust to neutron stars, the ultimate state of compaction, in which individual subatomic particles were forced together to form a conglomerate of tightly packed neutrons. The implosion would be accompanied by an explosion, releasing matter and energy into interstellar space and witnessed as a supernova event. Supernovae were predicted to emit light equaling that of hundreds of millions of suns. The total visible radiation emitted was claimed to be comparable to that emitted by the Sun over a period of 10 million years, this energy being but a small fraction of the total energy radiated at all wavelengths. (We now realize that the latter part of this claim is incorrect.) Material ejected from supernovae was estimated to have been a significant fraction of the mass of the initial collapsing star and to be traveling at velocities initially on the order of 15,000 to 30,000 kilometers per second. Supernovae were proposed as the sources of the mysterious cosmic rays impinging on the top of the Earth's atmosphere from outer space.

When the ideas of Zwicky and Baade were announced in 1933, the editor of *Science Newsletter* wrote, "Scientists are not inclined to accept the new theory until it shows its ability to withstand unfriendly criticism." And criticisms there were. Many of the leading scientists of the day, such as Eddington, tried to convince them that implosions played no role whatever in the evolution of the universe, and that white dwarfs represented the final stages of the evolution of all stars. Nobody of any stature in astronomy gave the supernova cosmic rays origin any credence, and the concept of neutron stars was ridiculed by some scientists for almost four decades, until the discovery of pulsars. Those four decades were to see a concerted effort by the avant-garde of California to prove the predictions correct. It is a remarkable testimony to the directed intuition of Zwicky and Baade that so many of the ideas they put forward 50 years ago are now accepted as correct, or mostly so.

Zwicky started his systematic survey for supernovae in 1934, using a 3.5-inch camera to photograph the rich Virgo cluster of galaxies from the roof of the Robinson astrophysics building of the California Institute of Technology at Pasadena. The experiment failed to discover a single supernova over a two-year period, although Zwicky had predicted the occurrence of several in that interval. The project caused a certain amount of amusement among some of Zwicky's Cal Tech colleagues who were eager to see their controversial friend and his absurd predictions proved wrong. Then in September 1936 the 18-inch Schmidt telescope was put into operation, and Zwicky was determined to "show the professional astronomers what a determined physicist can do. Beating the 'tar' out of the sky I found my first supernova in the spiral galaxy NGC 4157 in March 1937. On August 26, 1937, I discovered my second supernova in the dwarf spiral galaxy IC 4182 [object 4182 in the Index Catalogue] which has so far remained as the brightest one discovered in this century. A third excellent one I found on September 9, 1937, at the end of the same observing period, in NGC 1003."

These successes, plus a proposed whole-northern-sky survey program, encouraged Hale to ask the Rockefeller Foundation for an additional half-million dollars for the construction of a 48-inch Schmidt telescope, though because of the interruption of World War II this was not to be commissioned until 1949. The 18-inch Schmidt supernova survey had also suffered interruption, but it was restarted after the war, while the 48-inch Schmidt telescope was used for nine years exclusively for the Palomar Observatory All-Sky Survey, the standard atlas of the northern heavens. In 1959, a new large-scale supernova search was begun with the 48-inch Schmidt and was continued until 1976, when a limited survey was taken over again by the 18-inch Schmidt. Over 450 bona fide supernovae have been discovered over the past 90 years from many observatories around the world, and almost 75 percent of these have been discovered by the "Palomar pundits." Zwicky's personal score reached 120 shortly before his death.

The techniques for finding supernovae vary. Usually, selected fields of the sky are photographed, month by month, and the photographic films or plates compared for the appearance of new stars. With the 18-inch Schmidt survey at Palomar, films taken at different times were superposed and viewed under small-power binoculars to detect differences. An alternative technique was to superpose respectively a positive and negative of films spaced sufficiently in time and to project the ensemble—the supernovae then appearing as white spots on the projection screen. Another common method for detecting differences is to use a blink microscope, which switches rapidly between two photographic plates of galaxies. More recently, reference images of galaxies have been stored in computers and displayed alongside current exposures for comparison on a television screen. Advanced measuring machines under computer control are now being used to search photographic plates for faint supernovae. This technique may suffer, however, from the logistics problem of the sophisticated measuring machines not being sited at the observatories where the photographic plates have been obtained. An elaborate automated telescope system at Socorro, New Mexico, built for quickly scanning bright galaxies and processing the signals in a computer to permit immediate detection, failed to find a single supernova because of a host of problems bound to plague any new technical innovation. A new, advanced automated system has been developed at Berkeley, California, and plans are being made to use the Space Telescope to look deep into the cosmos to detect much fainter supernovae than could ever be seen from the ground. Until these new techniques have proved themselves, however, it seems that there is no better supernova detector than the human eye, laboriously scanning displayed images of galaxies.

If a suspected supernova is found on a plate, the galaxy is usually checked again for confirmation. No astronomer wants to be embarrassed by claiming a supernova discovery that was nothing more than a defect in photographic emulsion or a grain of dust on the plate. Once a discovery is certain, the usual procedure is for the discoverer to assess the supernova's brightness and position and inform interested observers at other observatories via a system of telegrams circulated by the Central Bureau for Astronomical Telegrams. Thereafter, a host of telescopes can be expected to be trained on the supernova, to record its change of brightness over a period of time and to analyze its light spectroscopically.

Extragalactic supernova surveys can provide information on how often and where in a galaxy supernovae occur. The distance to the parent galaxy can be determined by the technique outlined in Appendix 13. If a hypothetical superstar observer, blessed with eternal life, were living on a planetary system in a nearby galaxy so positioned that he or she could view the spiral structure of the Milky Way face on, that person would then witness the Galactic supernovae as sporadic, short-lived, extremely bright spots appearing randomly in time and space over the face of the Milky Way. At maximum, a supernova would outshine all the other stars in the Galaxy combined. The observer would have no way of predicting when the next bright spot would appear or even where it might appear. If, however, the person studied the distribution and frequency of the randomly occurring bright spots over a long enough period of time (say a few thousand years), he or she would soon be able to derive a characteristic interval between spots and predict where they were most likely to occur. This is just the sort of exercise astronomers on Earth have conducted in their searches for supernovae in external galaxies. Having to cope with a life-span of a mere three-score years and ten, however, an Earth-bound observer would find it futile to concentrate on a single galaxy, perhaps not witnessing a single supernova in that galaxy in a lifetime. Certainly one could not concentrate on just the Milky Way, since the dust lying along the Galactic plane would hide all but nearby supernovae from view. Observers have therefore concentrated their efforts on thousands of galaxies and infer from statistical analysis the characteristic intervals and spatial distributions for supernovae in different galaxy types.

For a loosely wound spiral galaxy of the type we believe the Milky Way to be (known in astronomical jargon as an Sc galaxy), a characteristic interval of about 20 years has been found. Our extragalactic observer colleague would presumably have been fortunate enough to derive a similar characteristic interval from patient vigil of the Milky Way and might also have concluded, as Earth-bound astronomers have done, that supernovae appear to be of two main types, occurring in Sc galaxies in about equal numbers. No doubt, he or she would have realized that supernovae of one of these groups, the Type II supernovae, are concentrated within the spiral arms of the Galaxy and occur with almost equal likelihood as one moves out from the central nucleus to beyond the solar neighborhood.

The easiest way to classify supernovae is to look at their light curves. The system of measuring stellar brightness, or magnitude, is derived from that adopted by Hipparchus in his original star catalog. Hipparchus used a simple scale of one to six to classify brightness, one being used for the brightest stars. This system has now been quantified as follows: negative numbers indicate a greater brightness than positive numbers, and each magnitude is 2.5 times brighter than the one next to it. The observed magnitude of a star is called its apparent magnitude. For example, zero apparent magnitude is the brightness of Vega, and −4 apparent magnitude is the brightness of Venus. To compare one star with another, we need to know their intrinsic magnitudes, such as would be obtained if the stars were placed at a standard distance. The distance chosen is 10 parsec—the parsec is an astronomical distance unit equal to 3.26 light-years. The magnitude of a star calculated as if it were at such a distance is termed its absolute magnitude. The absolute magnitude of the Sun is +5, although its extreme apparent magnitude because of its proximity is −26.

Apart from two main kinds of supernovae—Type I and Type II—other very rare types may exist (Zwicky used a classification scheme of Types I to VI), but since there remains some uncertainty as to the classification of these others, we shall only consider Types I and II here. Observations through standard filters—for example, the B filter for blue light—are used to compile light curves. The light curve for a Type I supernova shows a rapid rise, typically with absolute magnitude at a maximum of −19. After an initial drop of about three magnitudes in 20 to 30 days, the light curve then shows about a one magnitude fall in each successive interval of 50 days. Type II supernovae are rather more individualistic, although a common feature of their light curves is a plateau after the maximum, followed by a rather rapid decline. Their absolute magnitude near the maximum is usually −17 to −18, somewhat fainter than that of Type I. Light curves from typical Type I and II supernovae are compared in Figure 13.

Zwicky had originally concluded that there was only one type of supernova, since the first 12 he discovered were so similar (now known to be Type I). But then the next two supernovae discovered at Palomar were entirely different (the kind now known as Type II). The difference was recognized initially not so much from the light curves but rather from spectroscopic analysis of the supernovae's light.

Much of the pioneering work on the spectroscopy of supernovae was done by Rudolph Minkowski. His spectral data on the eighth-magnitude supernova discovered by Zwicky in

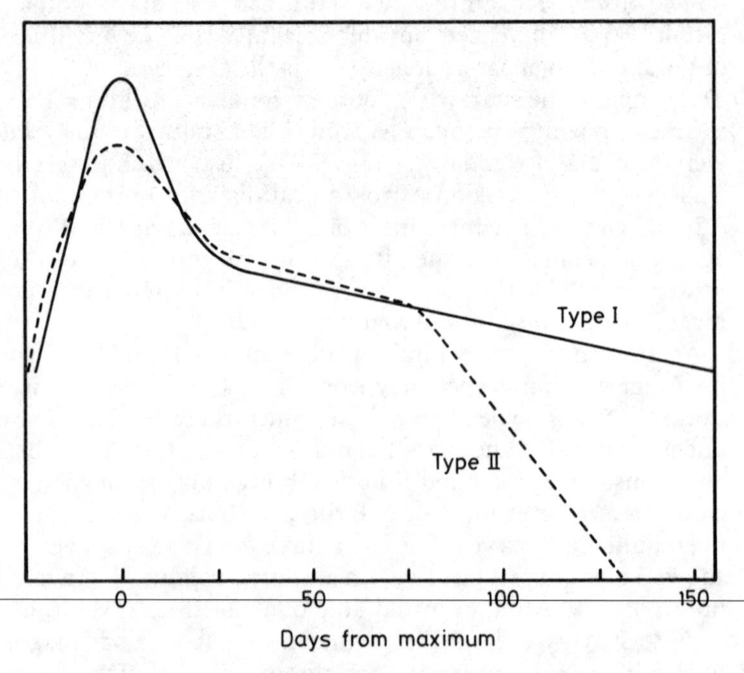

Figure 13. Light curves showing the typical variation with time of the light intensity from Type I and Type II supernovae.

IC 4182 remained the standard work on Type I spectroscopy until recent years. The 10 supernovae discovered after that in IC 4182 showed similar spectral properties, but Minkowski then made the important discovery that the next two did not. It was Minkowski who introduced the name Type I for the first group and Type II for those of "not Type I." It is amusing to note that Zwicky did not personally discover a Type II event until his thirty-sixth supernova.

The spectra of supernovae, shown in Figure 14, are found to be a complex mixture of continuum, emission, and absorption

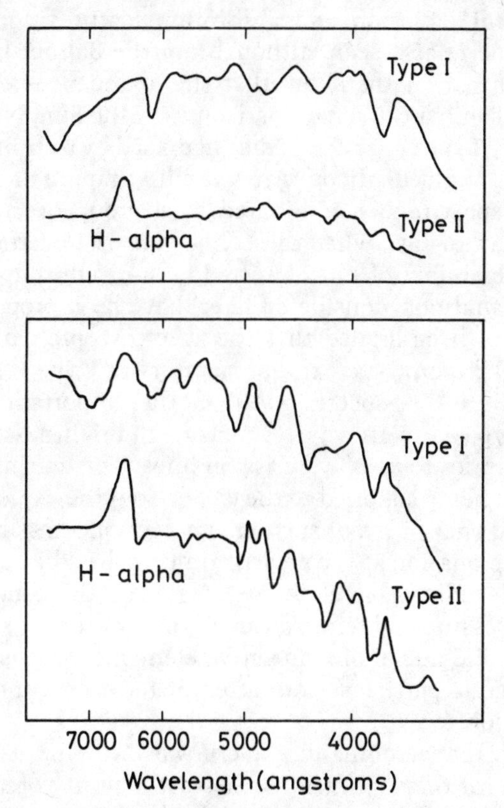

Figure 14. Sample spectra of supernovae. (Top) Near maximum brightness. (Bottom) About one month later.

lines. The clearly distinguishing feature of Type II's is that their spectra show strong emission lines of hydrogen (the so-called Balmer series of hydrogen lines) during the first few weeks following maximum light. In addition, there are strong emission and absorption lines from ionized calcium and un-ionized sodium, magnesium, and iron. The strength of the lines indicates that the relative abundance of these elements in Type II supernovae is not very different from their abundance in the Sun. The spectra of Type I supernovae are very much more complex. Initially the spectrum is dominated by a continuum, with several emission and absorption features common to Type II spectra also seen, although not the Balmer lines. The relative line strengths show that the abundances of various elements is enhanced in comparison with the Sun. Near maximum light, Type I spectra are distinguished by a strong absorption line of ionized silicon. After the first rapid fall in brightness, the spectrum is dominated by a group of four strong emission lines. Just what these lines are is uncertain. They may be a blending of lines emitted by ionized iron, although other explanations (considered later) have been proposed.

Later it will be argued that the spectroscopic data help to show what type of stars the progenitors of Type I and II supernovae were. The spectra reveal another important property. While a Type I supernova is decreasing in brightness, its spectrum continues to display emission lines. This means that energy is still being supplied to the ejecta from the explosion. But what could the source of such energy be? One possibility, the favored explanation at the present time, is that the energy may come from radioactive elements created in the conditions of extreme pressure and temperature at the time of the supernova outburst. (The nuclei of radioactive elements are unstable and emit energetic particles in an attempt to reach a more stable configuration.)

So much for searching for supernovae. Now let us consider the search for their remnants. The vast majority of supernova remnants identified in the Milky Way have been discovered at radio wavelengths. We have seen that radio supernova remnants are distinguished by the synchrotron nature of their emission with its nonthermal spectrum and evidence of polarization. Recognition of a nonthermal spectrum is not sufficient reason for classifying a radio source as a Galactic supernova remnant, since extragalactic radio sources also display this characteristic. In some cases, a definite distance estimate could establish that an extended nonthermal source lay within the Galaxy and must therefore be a supernova remnant. It might also be possible to identify optical filamentary structure and/or a historically recorded supernova for relatively nearby remnants, although of course the vast majority of radio remnants are believed to be so old that they probably went undetected and unrecorded. In the absence of any of the above evidence, the spatial distribution of the radio brightness of the source may support a supernova-remnant classification. From a study of many radio remnants, it is found that the majority show a characteristic ring structure, rather like a doughnut, often with a bite out of it. These ring or partial ring sources are indicative of shell-emitting regions centered on the sites of the original supernova outbursts.

Peripheral brightening, however, is not an essential property for radio supernova remnants. There appears to be a distinct class of radio remnant, the Crab Nebula being the best-known example, displaying an amorphous structure with central brightening usually attributed to continued injection of energy from an active pulsar at the center. The number of these plerionic remnants—so called from the Greek word meaning "filled center"—is difficult to estimate, since they tend to mimic ionized hydrogen regions in some of their properties. Eight have now been identified compared with the more than 100 identified supernova remnants displaying definite ring structure. More plerionic remnants undoubtedly await discovery.

Of five historically detected galactic supernovae in the past millennium, two (those of 1054 and 1181) left remnants showing central brightening, and on this basis it is tempting to speculate that perhaps as many as 40 percent of all supernovae produce plerionic remnants. A clue to the true nature of these strange objects may be the remnant MSH 15-56, shown in Fig-

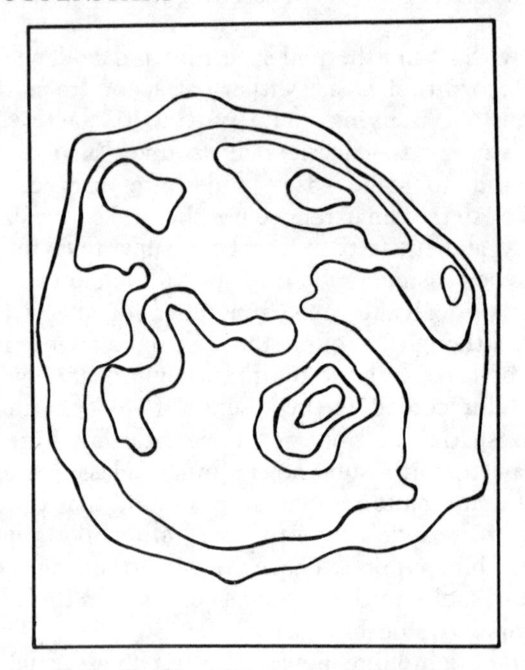

Figure 15. The radio remnant MSH 15-56.

ure 15, which has both a doughnutlike shell and a plerionic feature. We will speculate later that the plerions are produced by Type II supernovae.

The selection criteria described above have enabled scientists to classify some 130 extended Galactic radio sources as supernova remnants. The Galactic plane has now been very carefully surveyed so that present radio supernova remnant catalogs are believed to be nearly complete, at least for ring remnants.

During the late 1950s and the 1960s, the world saw the birth of giant radio telescopes. Some of the more spectacular examples were the 250-foot-diameter parabolic dish at Jodrell Bank, England, the 210-foot dish at Parkes in Australia, the giant Molonglo Cross (with arms a mile long) just 100 miles away, the 300-foot dish at Bonn, West Germany, and the largest of all, the 1,000-foot-diameter radio telescope built within a natural ravine at Arecibo, Puerto Rico. The large collecting area of these radio telescopes made them sensitive enough to detect faint emissions from radio remnants on the remote side of the Galaxy. Furthermore, the new radio telescopes were of higher resolution than earlier ones—that is, they were better able to resolve adjacent features and detect the characteristic ring structure of the majority of radio supernova remnants, distinguishing them from any nearby confusing sources. The new generation of radio telescopes uses an array of smaller parabolic dishes aligned so that the Earth's rotation causes them to sweep out and synthesize an aperture equivalent to that of a single giant radio telescope many miles in diameter. The remarkable resolution of the Earth-rotation aperture-synthesis radio telescopes has enabled radio photographs (radiographs) of supernova remnants to be obtained, revealing for the first time knots and fine structure in the shells of doughnut remnants and wisps and filaments in the plerionic remnants.

Aperture-synthesis techniques were pioneered in Cambridge, England, which now has a 5-kilometer-long rail-mounted system, in Holland with its famous Westerbork array, and in Australia. Although the northern hemisphere is now well-endowed with these sophisticated radio instruments, there is no synthesis array of comparable power in the southern hemisphere. Thus, a large portion of the Milky Way and such rich mines of astronomical knowledge as the Magellanic Clouds remain unexplored in the radio at very high resolution. The largest of the aperture-synthesis radio telescopes is the Very Large Array in New Mexico; this has the power and resolution to detect the faint radio emission from supernovae in distant galaxies and is being used to survey for the remnants of ancient extragalactic explosions as well as the faint radio echo following recent extragalactic outbursts.

Observational data from supernova remnants are now of such quality that for the first time detailed comparisons with theoretical models are becoming possible. The evolutionary behavior of radio remnants is at last starting to be understood, as are the distribution of radio supernova remnants throughout the Galaxy and the frequency of supernova outbursts of the type that leave long-lived radio remnants. Detailed studies reveal that supernova remnants are concentrated to within about 200 light-years of the Galactic plane and to within little more than the solar distance from the Galactic center. This probably does not reflect the true distribution of supernovae themselves, since a remnant's environment—that is, the interstellar magnetic field and background cosmic-ray particles—plays a dominant role in its radio appearance, at least in later life. The properties and distribution of the Galactic radio supernova remnants suggest that they are being produced no more often than about once every 80 years. This estimate is clearly at variance with the earlier estimate for the characteristic time interval between supernovae in Sc galaxies like our own of perhaps only about 20 years, but the discrepancy is resolved if one assumes that only about 25 percent of supernovae produce long-lived, ring-type radio remnants. Presumably, all events will initially brighten in the radio, most of them fading from view after a few hundred years. Only those 25 percent or so of supernovae occurring in suitable environments are likely to produce remnants detectable in the radio throughout their lifetimes.

There is a particularly important observation that can be made at radio wavelengths. Hydrogen, the dominant component of the universe, may emit radio waves of a wavelength of 21 centimeters. The nature of this emission is described in Appendix 14. Twenty-one-centimeter observations allow the distances to supernova remnants to be estimated.

Only in the radio can the remnants of supernovae be observed to the remote extremities of the Galaxy. The improved resolution of the new generation of radio telescopes is revealing structural and temporal details which will help to test and clarify present theoretical models of the evolution of supernova remnants and the energetics of supernova explosions. Radio data are also being used to understand the emission processes from supernova remnants. One can therefore confidently predict that radio observations will continue to contribute very significantly to our knowledge of these fascinating and spectacular stellar outbursts.

Pulsars are most readily identified in the radio. Only two—

in the center of the Crab Nebula and in the Vela remnant—have so far been identified optically. But over 300 radio pulsars have now been cataloged since the first chance detection in 1967 with a radio telescope at Cambridge, England, constructed for a completely different purpose. While Zwicky and Baade had predicted the formation of neutron stars, there had been no suggestion that they might be identified by pulsed radiation, which is why the discovery of the first pulsar caused considerable speculation as to its origin. Discovery of a second pulsar shortly after the first quickly dispelled thoughts of beamed messages from extraterrestrial civilizations, and rapidly rotating neutron stars were soon put forward as the most likely source. After the announcement of the first discovery, many of the giant, sensitive radio telescopes mentioned above were quickly assigned to survey for new detections. Only one radio pulsar is known to be a component of a binary system, and only three so far have been definitely associated with extended supernova remnants—the Crab Nebula, the Vela remnant, and an object called MSH 15-52.

A wealth of useful information can be obtained from spectroscopic surveys of the optical supernova remnants. In young remnants, where material ejected from the explosion can be expected to be identified, spectra may provide evidence for the synthesis of heavy elements in supernova explosions. In old remnants, where the ejecta have dispersed, the spectra should give information on the composition of the cooling shock-heated cloudlets in the interstellar medium. An example of the former situation is the young remnant Cassiopeia A (the strongest radio source in the Galaxy, which although not detected historically is believed to be no more than 300 years old). In Cassiopeia A oxygen, sulfur, and argon seem to have a much higher abundance with respect to hydrogen than is usual. Such anomalies tell us something about the progenitor star and the nuclear processes that precipitated the supernova outburst.

Although the interstellar medium consists mainly of hydrogen and helium (the two elements created in the Big Bang), significant traces of heavier elements created in the stars and fed to interstellar space by supernovae are present. Remnant spectra typically display emission lines characteristic of several elements: ionized hydrogen; singly ionized nitrogen (nitrogen which has lost a single electron) and neutral nitrogen (in which electrons are excited to higher energy states and then emit light of characteristic wavelengths when returning to lower energy levels); singly ionized, doubly ionized, and neutral oxygen; singly ionized iron; singly ionized and neutral helium; plus other minor components. The relative intensity of selected emission lines makes it possible, with careful modeling, to estimate the density, temperature, and composition of the emitting material. Since for many remnants the emitting material is believed to be shock-heated cloudlets, supernova remnants are useful probes of interstellar space and allow the composition and density of the interstellar medium to be studied throughout the solar neighborhood. Because of the effects of obscuration, most of the Galactic optical supernova remnants detected lie within 10,000 light-years of the Sun. About 20 optical remnants can be seen in the Magellanic Clouds, and a few tens in nearby spirals such as M33 and Andromeda. Doppler velocity estimates from optical remnant spectra make it possible to determine the velocity of expansion of optical remnants, providing valuable information on their dynamical evolution (see Figure 16).

A characteristic feature of many optical supernova remnants appears to be that the ratio of the intensity of two closely spaced red emission lines of singly ionized sulfur to the intensity of H-alpha emission is very much greater in remnants than in ionized hydrogen clouds. The reason for this is not that sulfur is more abundant in remnants, but rather that in ionized clouds most of the sulfur is in the doubly ionized state. This difference between the two types of nebulosity has been used as one of the standard techniques for identifying at least a subclass of optical supernova remnants. However, two other subclasses exist—one includes remnants showing hydrogen emission *only* and the other includes remnants displaying *no* hydrogen emission. It is likely that the latter subclass represents the debris from massive stars that have shed their hydrogen envelopes before exploding.

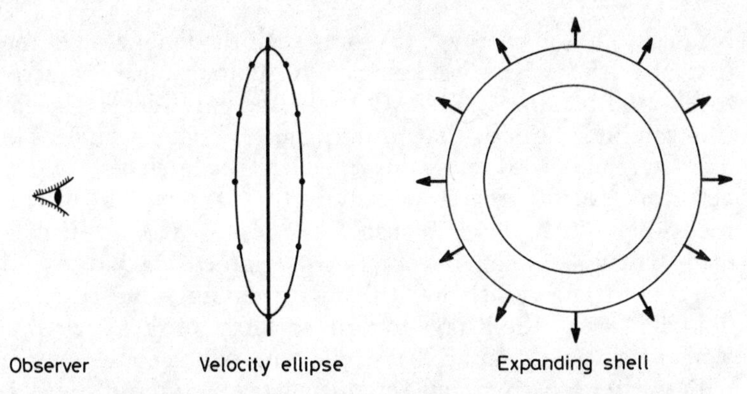

Figure 16. Estimating the velocity of expansion of a supernova remnant.

X-ray observations of extended supernova remnants have increased dramatically in recent years. Early generation X-ray instruments were designed to survey the skies and detect any X-ray source, regardless of its nature. It was realized that the vast majority of X-ray sources in the Galaxy were binary systems containing neutron stars or black holes with energetic X rays originating in the material from a companion star heated by accretion onto the compact object. Extended X-ray supernova remnants radiate mainly at low X-ray energies, and since the instrumentation to detect low-energy X rays is more difficult to construct than that for higher-energy X rays, plus the fact that low-energy X rays suffer more from obscuration than do high-energy X rays, early instrumentation on rockets and satellites was biased against the discovery of extended supernova remnants. Despite this, one of the first X-ray sources discovered was the Crab Nebula, which is one of the brightest X-ray objects in the sky.

The early detections of X-ray supernova remnants from rocket experiments occurred by chance. The first full survey of the X-ray sky was completed by the National Aeronautics and Space Administration (NASA) satellite *Uhuru,* launched in 1970. But since this experiment failed to detect the low-energy X rays in which extended supernova remnants were expected to radiate most brightly, few new discoveries of X-ray remnants were made. The first opportunity to survey the complete

X-ray sky in low-energy X rays was with the launching of the first of NASA's HEAO (High Energy Astrophysical Observatory) satellites in 1977. HEAO-1 was one of the largest scientific satellites built at that time: 6 meters high, 2.4 meters in diameter, and 2,700 kilograms in weight. Its low-energy X-ray detectors had far greater sensitivity than any similar instruments previously flown in space. As a result, HEAO-1 discovered 10 new X-ray supernova remnants in the Galaxy to add to the list of 10 previously identified with certainty. But HEAO-1 did not "focus" the X rays to form an image of the X-ray sky. What was needed was an X-ray telescope, which could operate with greater sensitivity and produce X-ray pictures of the sky. This was achieved with the launch in 1978 of NASA's follow-up mission, HEAO-2. In orbit, HEAO-2 was christened "The *Einstein* Observatory" to commemorate the centenary of the birth of the twentieth century's greatest scientist. It has been claimed that the *Einstein* X-ray telescope represented as great an advance in sensitivity over the first X-ray detectors carried above the atmosphere by rockets less than two decades earlier as the 200-inch optical telescope on Mount Palomar represented over Galileo's first telescope of more than three centuries earlier—such has been the accelerating pace of astronomical technology.

The advances in our understanding of X-ray supernova remnants following the *Einstein* observations have been truly monumental. For the first time, astronomers have high-resolution X-ray images of supernova remnants, showing the complicated web of interwoven hot-gas filaments formed by the expanding shock. And X-ray spectroscopy of acceptable resolution has been achieved for the first time. X-ray remnants have been cataloged for the Magellanic Clouds and detected in other galaxies. Astrophysicists will need time to interpret the gold mine of data from *Einstein,* but our view of the heavens has been changed in dramatic fashion by this remarkably successful space mission.

While the early history of X-ray astronomy was dominated by spectacular discoveries about neutron stars and black holes in binary systems, the recent history has included great advances in the observations of extended X-ray remnants. The X-ray remnants of supernovae appear to be among the most valuable of astronomical antiquities, capable of revealing information on the initial supernova outburst, the evolution of the remnants, and the structure and composition of the interstellar medium.

The high-quality observational data at X-ray, optical, and radio wavelengths has stimulated renewed theoretical studies of how supernova remnants evolve (see Appendix 15). Of particular importance is knowing the age of a remnant, and this brings us back to the historical data. From the records of new stars it should be possible to decide whether a new star was a nova or a supernova. In the case of a supernova, historical descriptions should make it possible to determine an accurate enough position to identify the remnant of the outburst with certainty. Such identifications then make it possible to study how supernova remnants evolve with time and to test theoretical models and predictions. The importance of this task was noted in 1959 by the distinguished historian of Chinese science, Joseph Needham:

> The extent to which the Chinese records of guest stars remain of living interest to current astronomical research may be seen in the field of radioastronomy, where during the past few years great additions to knowledge have been made. . . . The rapid upsurge of this new and powerful method of study of the birth and death of stars . . . makes urgently necessary the reduction of the information contained in the ancient and medieval Chinese texts to a form utilisable by modern astronomers in all lands. For this purpose, however, collaboration between competent sinologists and practical astronomers and radio astronomers is indispensable.

And Fritz Zwicky noted: "The investigation of the remnants of supernovae and their relation to historical records, both written and unwritten, will be one of the most fascinating tasks awaiting the next generation of astronomers. . . ." Since almost all the pre-Renaissance records of new stars originate from China, Korea, and Japan, let us return to the superstar surveys of the ancient Orient.

5

SUPERSTARS AND PORTENTS

> Probably another reason why many Europeans consider the Chinese such barbarians is on account of the support they give to their astronomers—people regarded by our cultivated Western mortals as completely useless. Yet there they rank with Heads of Department and Secretaries of State. What frightful barbarism!
>
> *Franz Kühnert (1888)*

THE GREAT RICHNESS of astronomical data from the ancient Orient results in part from the official character of Chinese astronomy. By the time of the Han dynasty (202 B.C.–A.D. 220), an astronomical office had been established as a special subdepartment within the Ministry of State Sacrifices. Throughout subsequent Chinese history, as well as later in Korea and Japan, and even until modern times, the Astronomical Bureau existed as an important government office. It should be remembered that the ancient Chinese were the first civilization to introduce a central bureaucracy—a civil service with entrance by written examination. The leading role astronomers played in the Chinese official hierachy is stressed in the following imperial edict issued in 840 during the T'ang dynasty:

> If we hear of any contact between the astronomical officials, or their subordinates, and officials of any other government department, or miscellaneous common people, it will be regarded as a violation of security regulations which should be strictly adhered to. From now onwards, therefore, the astronomical officials are on no account to mix with civil servants and common people in general. Let the censorate see to it.

There are no doubt many professional astronomers today who would enjoy the official esteem and support given to their ancient Oriental counterparts.

The Astronomical Bureau had two main functions. The first was to maintain an accurate calendar, an important consideration for a people so dependent on a stable agricultural system. For this reason, Chu Yüan-chang, the founder of the Ming dynasty, established his own Astronomical Bureau in 1367. The second function of the bureau was to observe and interpret celestial portents—hence the large number of records of phenomena, such as new stars, of no calendrical importance. The belief in astral influence on state events seems to have germinated very early in Chinese history, so that by the Han dynasty an elaborate system of political astrology had developed. An imperial observatory was built at each of the various capitals of China. When the capital was moved, as after the fall of a dynasty, a new observatory was built. For later dynasties, suspicion that subordinates could manipulate astrological data to their own purpose led the emperors to establish independent observatories within the confines of their palaces. It seems this precaution was to little purpose. P'eng Ch'eng, who was Astronomer Royal in China between 1068 and 1077, remarked:

> In our [Sung] dynasty, the Astronomical Department [of the imperial academy] was established within the Imperial Palace, with a clepsydra [a water clock], and an observatory with a bronze armillary sphere, just the same as in the Astronomical Bureau. The reports from the two observatories are supposed to be compared. Every night the Astronomical Department should state whether or not there have been vapors, unusual auspicious phenomena, oppositions and conjunctions, changes of position of the heavenly bodies, etc., and

> present its report before the opening of the palace gates. Then when the gates have been opened [at dawn], the results of the Astronomical Bureau should arrive, after which the two should be compared and checked, in order to avoid all false reports. But in recent times the officials of the two observatories have secretly copied from each other before reporting, and this went on for years. Everyone knew about it, yet no one thought it strange.

Even today astronomers are sometimes suspected of having copied the results of others, without due acknowledgment—evidently a practice of long standing.

More than six centuries and three dynasties later, in 1698, the Jesuit Louis Lecomte described the scene as he saw it in Peking in similar terms:

> Five astronomers spend every night on the tower in watching what passes overhead; one is gazing towards the zenith, another to the east, a third to the west, the fourth turns his eyes southwards, and a fifth northwards, that nothing of what happens in the four corners of the world may escape their diligent observation. They take notice of the winds, the rain, the air, of unusual phenomena, such as eclipses, the conjunction or opposition of planets, fires, meteors, and all that may be useful.

Another important department of the Chinese civil service was the Bureau of Historiography. It became standard practice from Han times for the Bureau of Historiography to compile an official dynastic history for the preceding dynasty. The compilers were given free access to all official records, including those of the Astronomical Bureau. So strong was the ancient Chinese belief in the importance of history that the dynastic histories are believed to represent reliable accounts of the main state events from the deposed dynasties, although of course the written accounts merely reflect what the official historians saw fit to preserve. In particular, the astronomical records, which were usually contained in separate astronomical treatises, are just summaries of what presumably were very detailed firsthand accounts. The dynastic histories represent an almost continuous record covering nearly 2,000 years, are without equal from any other civilization, and offer a treasure trove of useful astronomical data.

In these astrological records, some attempt was usually made to describe where transient phenomena such as new stars appeared. The sky was subdivided into 28 "lunar mansions" lying roughly on a great circle near the celestial equator. The lunar mansions were used in positional astronomy for expressing "right ascension," the celestial equivalent to longitude. Some 250 to 300 asterisms, or small constellations, were then identified, mostly containing no more than five or six stars visible to the naked eye. All these star groups and some individual stars were associated with certain affairs on Earth. Typical names of asterisms were Emperor, Crown Prince, the Nobility, Minister of Works, Celestial Temple, Outer Kitchen, Celestial Dog, Father-in-law, Army Wells, Celestial Orchard, Weeping Star, Celestial Coin, Undertakers, Court Eunuchs, Celestial Arsenal, Guest Houses.

A celestial event, such as the appearance of a comet or a new star in a particular asterism, was regarded as a precursor of terrestrial events. There was then a standard prognostication for every omen. For example, "When a comet guards the stars called the Supervisors of the Nobility, while their colour also fades, they predict that these officials will be busily engaged. When a new star guards them, changes of ranks and titles are to be anticipated." Clearly such standard prognostications might be expected to have been self-fulfilling. Furthermore, the delay between the appearance of an omen and its acknowledged fulfillment tended to be conveniently long. The solar eclipse of A.D. 120, which was almost total at the Chinese capital of Loyang, was regarded as boding ill for the dowager empress. It was not until two years and three months later that the portent was fulfilled—the dowager empress died. A delay of that length was typical, evidently to ensure the maximum chance of success. Even as late as 1882, the following could be written in the official annals of the reign of Emperor Kuang Hsü, the same reign during which the Boxer Rebellion occurred:

> When a tailed comet was seen last year, an imperial edict was written to the palace and court officials, ordering them

> to perform their respective duties conscientiously. In the second decade of this month the comet was seen again in the southeast. This must be due to the frequent mistakes committed by those employed in the administration; the hardship of the village people had not been adequately presented to the throne. Order is now given to make a thorough investigation.

The fact that astrology was officially condoned as recently as this emphasizes just how static Oriental astronomy had remained at a time of major scientific advances in the West.

The prognostications accompanying astronomical observations may seem trivial now, but it is as a direct result of the importance attached to astrology that we possess such an impressive list of new star sightings. And since the substance of a particular prognostication depended on which asterism the new star appeared in, records usually give some idea of location.

Although astronomy was under political control and had a strong astrological orientation, there is little evidence of fabrication or falsification. We can now calculate, with the help of computers and a detailed knowledge of the motion of Sun, Earth, and Moon, whether an eclipse or occultation could in fact have occurred. Similarly, descriptions of planetary behavior can be checked. The ancient Oriental observations seem to have been remarkably accurate for naked-eye observations and are basically reliable. Very few were blatantly fabricated records. A rather amusing example is from a Korean chronicle covering the year 1406:

> August 10. The Moon invaded Mars
> August 14. The Moon invaded Mars
> August 17. The Moon invaded Mars; Jupiter and Mars invaded one another
> August 21. Mars invaded Jupiter
> August 25. The Moon invaded Jupiter

The term for *invaded* was used to describe two astronomical objects coming into close proximity, within about a one-degree separation. This account, which reads like a fifteenth-century version of *Star Wars,* is almost entirely contrived. Computation shows that only the events for August 21 and 25 occurred. Although such glaring fabrication was rare, there is evidence that during times of political turmoil there was an increase in the number of phenomena of astrological significance recorded. It seems that during periods when there was apparently little need for political or social prognostications, trivial celestial events may not have been recorded. It is highly unlikely, however, that anything as spectacular as a new star would have escaped the attention of either astronomers or historians.

The records show that three kinds of new stars were recognized in the Far East. The first were known as *k'o-hsing*—"guest stars" or "visiting stars." The well-known new stars of 1006, 1054, 1572, and 1604 were all described as *k'o-hsing,* so we might expect the term to be synonymous with novae and supernovae. Occasionally, however, the term is used where there is also allusion to motion, indicating that the object in question was in fact a comet. So, while one expects to find novae and supernovae described as *k'o-hsing,* there is a need for caution. The standard term for a comet with discernible tail (the second category of new star) was *hui-hsing*—"broom stars" or "sweeping stars." In the official history of the Chin dynasty (A.D. 265–420) the following description of a *hui-hsing* is given: "Its body is a sort of a comet, while the tail resembles a broom. Small comets measure several inches in length, but the larger ones extend across the entire heavens." The final kind of new star categorized in Oriental annals were the *po-hsing*—"rayed stars" or "bushy stars." The term seems to have been used to describe an apparently tailless comet. Again in the history of the Chin dynasty, we read: "By definition a comet pointing toward one particular direction is a *hui* comet, and one that sends its rays evenly in all directions is a *po* comet." Other miscellaneous terms were sometimes used—the 1006 supernova was called a *chou-po* star (an "Earl of Chou" star), a special classification apparently to emphasize its exceptional brilliance and auspicious nature. The term *ch'ang-hsing* ("long star") was sometimes used as an alternative to *hui-hsing.*

By careful searching in Oriental dynastic histories, encyclopedias, diaries, and astronomical works, investigators have produced a large number of new-star records. The Swedish astronomer Kurt Lundmark seems to have been the first astronomer in modern times to appreciate the value of this fascinating field of research, using it in his efforts to gather historical accounts of sudden appearances of superbright stars to support his hypothesis for the existence of events distinct from normal novae. Searches for historical new-star records had been made by Edouard Biot and Alexander von Humboldt in the mid-nineteenth century. Much of the inspiration for more recent surveys came from the monumental work on ancient Chinese astronomy of Joseph Needham and his student Ho Peng Yoke in the 1950s. Over the past few years, much noteworthy work on historical astronomical records has originated in the People's Republic of China.

Almost all historical records of *hui-hsing* and *po-hsing* mention motion, and we can therefore assume that these stars were not comets. This reduces the list of potential novae and supernovae from a full 2,000 years of records to just 75 events. A remarkable although perhaps not totally unexpected fact is that for these 75 cases, the records indicate that the new stars were seen for either fewer than 25 or more than about 50 days. Events of intermediate duration are completely absent. Because a conspicuous supernova of any type fades fairly slowly, with a naked-eye visibility of several months or more, we can ignore the short-duration stars, which were probably novae, and concentrate on the few long-lasting objects. If one considers only the *k'o-hsing* which lasted for more than fifty days, the list of supernova candidates is shortened to 20.

One further selection can be made. Modern studies of novae in our galaxy give an average distance to those bright enough to be detected with the naked eye as about 1,500 light-years. The average distance of novae from the plane of the Galaxy in the neighborhood of the Sun is about 1,000 light-years. We would therefore expect novae visible to the naked eye, and recorded historically, to be distributed more or less uniformly over the celestial sphere. By contrast, extragalactic surveys show that supernovae in spiral galaxies like our own tend to be concentrated in the plane of their parent galaxy. Because of the vast average distance from us of galactic supernovae that can be detected with the naked eye—the nearest historically observed supernova, that of 1006, was about 3,000 light-years from Earth—they would be expected to lie close to the plane of the Galaxy. By restricting our attention to the historical new stars lying, say, to within 25 degrees of the Galactic plane, we reduce our list of candidates to just 13 (see Figures 17 and 18). This

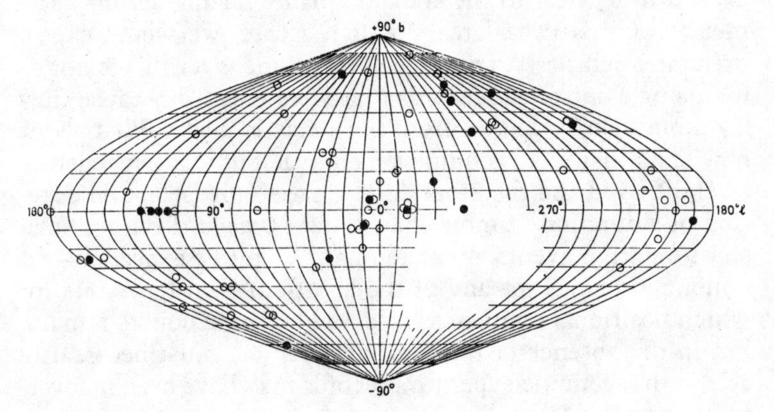

Figure 17. The Galactic distribution of pretelescopic novae and supernovae.

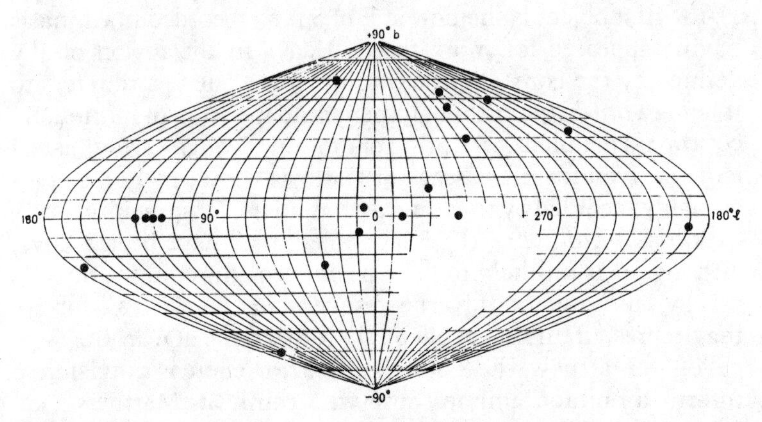

Figure 18. The Galactic distribution of pretelescopic new stars of long duration.

short list belies the enormous amount of research carried out over many years by a few dedicated individuals searching, researching, interpreting, and reinterpreting the historical literature.

The selection process described above poses the obvious danger that a genuine historical supernova record may have been rejected; might researchers have thrown out the baby with the bath water? What about a supernova so close to the Earth that even if it lay near to the Galactic plane, its angular displacement would still be large? In such a case, we would expect extreme brightness to have been mentioned, yet this is not so for a single one of the records that were rejected because they lay more than 25 degrees from the Galactic plane. What about new stars that were genuine supernovae but just happened to be short-lived and therefore did not satisfy the selection criterion of a duration of more than about 50 days? Even for such short-duration events, some remnant might be expected—yet none are obvious for any of the short-duration new stars for which positional estimates are possible. Some doubts remain, but in the absence of further evidence, one must necessarily accept that genuine supernova records may have been inadvertently discarded.

Of the 13 supernova candidates on our final short list (Table 1), the first object is the new star of 5 B.C., recorded in China as having appeared for more than 70 days in the region of sky defined by the constellation of Capricorn. The record refers to it being a *hui-hsing*, although there is no mention of motion. By contrast, a detailed description of motion through various asterisms is given for the passage of Halley's comet in 12 B.C., presumably recorded by the same astronomers. Despite its classification as a *hui-hsing*, the 5 B.C. object may have been a nova, although it is unlikely to have been a supernova since there is no obvious remnant. It has been suggested that the 5 B.C. object may represent an independent sighting of the star in the Gospel of St. Matthew, the Star of Bethlehem. There is a division of informed opinion among those who think St. Matthew's account is purely mythical, those who prefer to believe that the event was truly miraculous, and those who have sought an

TABLE 1

New Stars of Long Duration, Close to the Galactic Plane

Date	*Where Sighted*	*Duration*	*Angular Distance from Galactic Plane*	*Comments*
5 B.C.	China	70+ days	−25°	Nova/comet
A.D. 185	China	20 months	−2°	Supernova
369	China	5 months	Unknown	Case unsolved
386	China	3 months	0°	Possible nova
393	China	8 months	0°	Supernova
396	China	50+ days	−25°	Probable nova
1006	China, Japan, Europe, Arab lands	Several years	+15°	Supernova
1054	China, Japan, Arab lands	22 months	−5°	Supernova
1181	China, Japan	185 days	+3°	Supernova
1572	China, Korea, Europe	16 months	0°	Supernova
1592	Korea	3 months	0°	Suspect objects
1592	Korea	4 months	0°	(Suspect objects)
1604	China, Korea, Europe	12 months	+7°	Supernova

astronomical explanation. Astronomical explanations proposed include comets, novae, supernovae, and the close conjunction of planets. The gospel account fails to distinguish between these possibilities: "Some men who studied the stars came from the east to Jerusalem, and asked, 'Where is the baby born to be King of the Jews? We saw his star when it came up in the east, and we have come to worship him.'" Later the account reads: "and on their way they saw the same star they had seen in the east. When they saw it, how happy they were, what joy was theirs! It went ahead of them until it stopped over the place where the child was."

A literal interpretation of the account belies a possible astronomical explanation, since no astronomical phenomenon could move "ahead" (in this case south from Jerusalem to

Bethlehem—stars "move" east to west), and stop "over the place." Nevertheless, a number of astronomical events did occur about the time theologians and historians now believe Jesus was born. Contrary to tradition, this was not the twenty-fifth of December A.D. 1. Such a date is inconsistent with biblical chronology and was erroneously derived by a process of backdating in the sixth century. Four years of the reign of Emperor Augustus (when he reigned under his own name of Octavian) were missed in this calculation, putting the birth back to pre–4 B.C., such an early date also being required to have Jesus born (according to scripture) during the reign of King Herod. The death of Herod, just before the Passover of 4 B.C., can be accurately dated from an account of the Jewish historian Josephus and his reference to a lunar eclipse a few days before Herod's death. Joseph and Mary's journey to Bethlehem was necessitated by a census, and we know that a census was ordered by the Emperor Caesar Augustus in 8 B.C. A birth year between 8 B.C. and 4 B.C. therefore seems probable.

In the year 7 B.C. there was a conjunction of the planets Jupiter and Saturn. Although comparatively unspectacular (the planets were never separated by less than about one degree on the sky), the conjunction was unusual in that the planets appeared almost stationary with respect to the fixed stars, making three close approaches within a seven-month period (see Figure 19). It has been argued that this triple conjunction could have had special astrological significance, even though it does not fit the gospel description of a star. There were no known spectacular comets during the period of interest, Halley's comet having last transited in 12 B.C. There was of course the 5 B.C. new star in Capricorn during March and April. In 4 B.C. another new star is recorded, in the same region of sky (certainly within about 20 degrees of the 5 B.C. object's position). The positional proximity and same months of appearance have raised the question whether the second record was merely an allusion to the 5 B.C. event misplaced in the historical records by a year, especially since similar possible errors in date are known. Even if the 4 B.C. new star was a distinct event, there is again no compelling reason to believe it was a super-

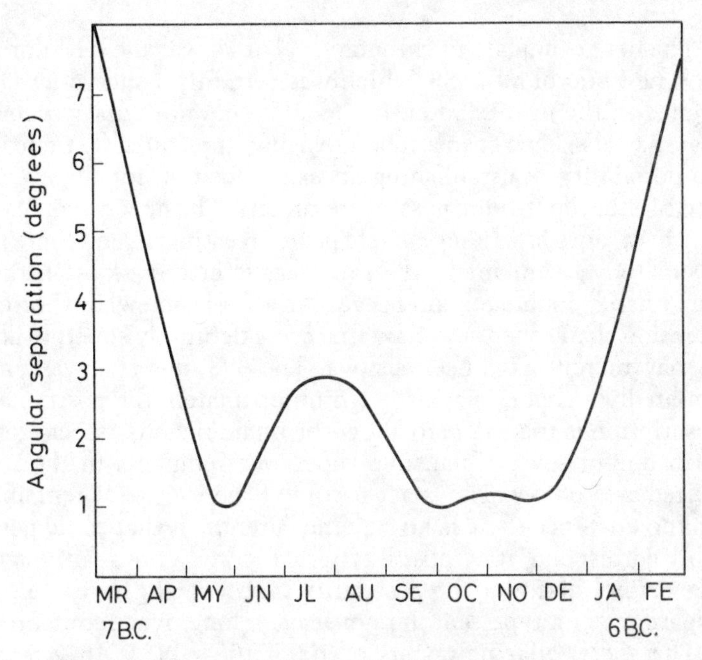

Figure 19. The angular separation of Jupiter and Saturn during the 7 B.C. triple conjunction.

nova, although there is a stellar remnant of a comparatively recent supernova in the same vague area of sky—the only known binary pulsar, mentioned earlier.

The new star of 5 B.C., and to a lesser extent that of 4 B.C. if it was a distinct event, fit the biblical story chronologically, and they were in the right part of the sky to be seen to "come up in the east" (a more exact astronomical translation might be that the star was seen "in the first rays of dawn"). So could either represent an independent sighting of the Star of Bethlehem? This question remains one that astronomers, theologians, historians, and Sinologists still actively debate, and it will undoubtedly remain a question never answered. Nevertheless, every year as Christmas approaches, a host of well-meaning but often misdirected researchers come forward to argue publicly for the theory they favor.

The next candidate in chronological order was the long-duration new star of A.D. 185, which was certainly a supernova, as argued in the next chapter. In 369 a five-month-duration new star may also have been a supernova, but unfortunately there is no possibility of establishing an exact position for the event and of thus identifying possible remnants. The new star of 386, however, does have a reasonable position estimate, and there is a supernova remnant in the same vague area of sky. On the other hand, doubts remain because it was of somewhat shorter duration than other new stars that were definitely supernovae. It may merely have been a nova. The 393 new star was undoubtedly a supernova, although unfortunately the positional description is too vague to allow the unambiguous association with one of several plausible supernova remnants in the required area of sky. The exact nature of the 396 event remains unknown. Since there is no supernova remnant that could possibly correspond, it is most likely that it was a nova, although one cannot discount the possibility that it might have been a supernova of a type that did not leave a long-lived remnant.

The spectacular new stars of 1006, 1054, 1181, 1572, and 1604 were all undoubtedly supernovae, for which definite or probable remnant identifications have been suggested. These supernovae will be discussed individually in the chapters to follow. This leaves just the two new stars sighted in Korea in 1592. A previous new star had also been recorded in Korea in 1592, on November 28, though it does not qualify for our short list since it was 70 degrees from the galactic plane. Far from being a new star, it was almost certainly a sighting of Mira, a bright variable star. Just two days later came the report of a new star close to the Galactic plane, followed four days afterward by yet another supposedly independent new star again close to the plane. This would mean that three new stars were seen within six days—and only by the Koreans. It seems unlikely that the diligent Chinese should not have recorded them, or the Europeans, who were now aware of new-star phenomena. It appears that the records of one or both the two new stars supposedly sighted in Korea close to the Galactic plane may be fabrications, despite the fact that the individual records themselves appear at first sight quite genuine. The case of the Korean new stars of 1592 remains unsolved.

One other new star, in 1408, is worthy of note. It did not make the short list given in Table 1 since the extant records give no idea of duration. Nevertheless, the records suggest that it must have been exceptionally bright, and a reasonably good description of its position is given. We will return later to the intriguing new star of 1408.

Undoubtedly, the historical new-star records are incomplete. Some records must have been lost. Certain new stars would have been too far south to be seen by northern hemisphere civilizations (particularly in the Orient) with their written histories. Some might have been daytime objects, which were not bright enough to be seen by day and which would have faded sufficiently so as not to be recognized as new stars when next they were nighttime objects. Perhaps certain supernovae are not intrinsically bright optically or lie in regions of very high obscuration. Are there, then, any young, nearby remnants for which historical sightings might have been made? The answer is that there are only three. One is the radio-bright young remnant Cassiopeia A, believed to be only about 300 years old. The others are southern remnants called RCW 103 and MSH 11-54, which were almost certainly created in supernovae within the past thousand years. RCW 103 and MSH 11-54 are too far south to have been seen from China. We will return later to the question of whether the supernova that gave birth to Cassiopeia A might in fact have been observed.

Seven definite supernovae, plus a couple of possibilities, may not seem a very good catch from such a large haul of historical astronomical records. When we remember, though, that Galactic supernovae are rare events, and only those on the near side of the Galaxy could possibly have been detected from Earth, the historical surveys do not appear so inadequate. Certainly the records that have been left have helped to reveal the true nature of the superstars and their remnants, as we shall see when we attempt to relate the historical supernovae to the remnants visible today.

6

THE WORLD ALIGHT

The burst of new light by its suddenness,
always appears inimical to the unprepared heart.

Jean Paul

THE NEW STAR of 185 is the earliest recorded for which there are any grounds for supposing it to be a supernova. The star was reported only in the Chinese annals, and all we know about it is gleaned from a single text. Since the star appeared in the constellation of Centaurus and was only visible south of about latitude 35° north, Europe cannot have been a source of observations, although possible vague allusions to the event in Roman literature were mentioned in Chapter 2. There seems a possibility that the star might have been carefully observed in Alexandria (31° north), the home just half a century earlier of the great astronomer Claudius Ptolemy and a mecca of observational astronomy at this time, but even if it was seen, no account of it has survived.

The solitary Chinese description of the new star is to be found in part of the astronomical treatise of the official history of the Later Han dynasty. A translation of the text is as follows:

> [A.D. 185, December 7] . . . a guest star appeared within Nan-mên [an asterism, the "southern gate"]. It was as large as half a mat; it was multicolored, and it scintillated. It grad-

> ually became smaller and disappeared in the 6th month of the year after next. According to the standard prognostication, this means insurrection. . . . [Four years later] the governor of the metropolitan region Yüan-shou punished and eliminated the middle officials. Wu-kuang attacked and killed Ho-miao, the general of the chariots and cavalry, and several thousand people were killed.

Emperor Hsiao-ling, who was on the throne when the guest star occurred, reigned from 168 to 189. His capital was Lo-yang (35° north), which had been the seat of government since the beginning of the Later Han dynasty. On the basis of the highly technical nature of many of the astronomical records in the treatises, and by inferring the position of most solar eclipse sightings, we can confidently assume that nearly all astronomical observations recorded in the dynastic history were made at Lo-yang by the imperial astronomers. This is the likely site of the recorded new-star observation.

It is quite possible that at various periods astronomical records were falsified for astrological purposes, and this might cast doubt on the unsubstantiated account of the 185 new star. At this epoch, however, the imperial astronomers appear to have been skilled observers, and in any case, as already mentioned, the reliability of their observations can be confirmed by checking particular planetary observations. The positions of the planets for any date can be calculated from our present knowledge of their orbits. If, for instance, a historical record refers to a particular date when a planet entered an asterism or was occulted by the Moon, or when two or more planets passed in close conjunction, then these claims can be assessed. There can be little doubt that only a minute portion of the original observations of this type have survived. During Hsiao-ling's reign, the dynastic history gives only five references to Venus, four to Mars, and one to Jupiter, and Saturn seems not to have been noticed at all.

Despite the apparent dilatoriness of Hsiao-ling's astronomers insofar as planetary phenomena are concerned, each reported observation was regarded as of very great omen value and was accompanied by a lengthy astrological commentary.

All the planetary records have been checked and found to be accurate, with the exception of minor errors in date. From this analysis, it is clear that there is no evidence whatever of deliberate falsification of astronomical observations during Hsiao-ling's reign. Numerous celestial events may have been omitted from the history, deliberately or otherwise, but those that are recorded are very reliable and allows us to place considerable confidence in the account of the 185 new star.

An inferred period of visibility as long as some 20 months makes it highly probable that the star was a supernova, while a comet can be entirely ruled out. A duration of more than a year implies that the text remains silent on the disappearance and recovery of the star around the time of conjunction with the Sun, but this is not a major problem. The supernova of 1054 is far better documented than the star of 185, and yet although both the dates of first sighting and final fading are given, corresponding to a period of visibility of nearly two years, there is not a single reference to heliacal setting or rising.

The asterism Nan-mên, in which the guest star appeared, was probably the southernmost star group visible from central China. In the astronomical treatise of the history of the Chin dynasty, we read: "The two stars of Nan-mên, situated south of K'u-lou, form the outer gate of the heavens and govern garrison troops." By comparing several ancient star charts, it is possible to identify the components of Nan-mên with some certainty as the bright stars Alpha and Beta Centauri, two of the ten brightest stars in the sky.

As for the position of the new star in relation to Nan-mên, it was described as "within" (chung) the asterism. The usual expression to denote the location of a planet or new star is the rather vague term *yu,* which may be translated "at" or "in the vicinity of." The word *chung,* however, which is comparatively rare, is more specific. Depending on the context, it can be rendered as "middle" or "within." It is used with the former meaning in the name for China itself, Chung-kuo (the Middle Kingdom). The latter meaning, on the other hand, is exemplified in the astronomical treatises: planets are said to enter "within" asterisms; occultations are described as planets entering "within" the Moon; sunspots are referred to as seeds or vapours "within" the Sun. *Chung* was a precise astronomical term, placing the 185 new star somewhere approximately between the two stars Alpha and Beta Centauri.

The new star could never have been far above the horizon at Lo-yang, certainly no more than a few degrees, and would only have been visible for a very few hours each day. At discovery it was rising just ahead of the Sun, so that its visibility must have been seriously impaired by the dawn glow. The fact that it was nevertheless observed as a bright object emphasizes its exceptional brilliance, and this is confirmed by its period of visibility, some 20 months. Stars are invisible to the naked eye when fainter than magnitude about +6. A typical supernova fades about 12 magnitudes in 20 months. Since the new star of 185 was close to the horizon, atmospheric extinction would certainly have reduced its brightness by at least two magnitudes. An apparent magnitude at maximum of at least −8 seems probable, rivaling the quarter Moon. Such a bright supernova must have been relatively nearby—certainly closer than about 5,000 light-years distant.

In looking for a possible remnant of the outburst, we are therefore restricted to remnants lying approximately between Alpha and Beta Centauri and closer than 5,000 light-years. There is one obvious such candidate, an extended radio/optical/X-ray source called RCW 86, showing the characteristic ring structure plus other properties of a supernova remnant (see Figure 20). This is just a few degrees south of the line joining the two stars of Nan-mên, and it would have been above the Lo-yang horizon in 185. The radio, X-ray, and optical properties of RCW 86 are all compatible with its being less than 2,000 years old and closer than 5,000 light-years to the Earth. One other object of interest, however, must be mentioned: lying exactly on the line between Alpha and Beta Centauri and therefore satisfying perfectly the description "within" is a planetary nebula, surrounded by a faint shell of high-velocity optical filaments (but not seen at radio or X-ray wavelengths) that appear to have been ejected from the vicinity of the planetary nebula about 2,000 years ago. If this high-

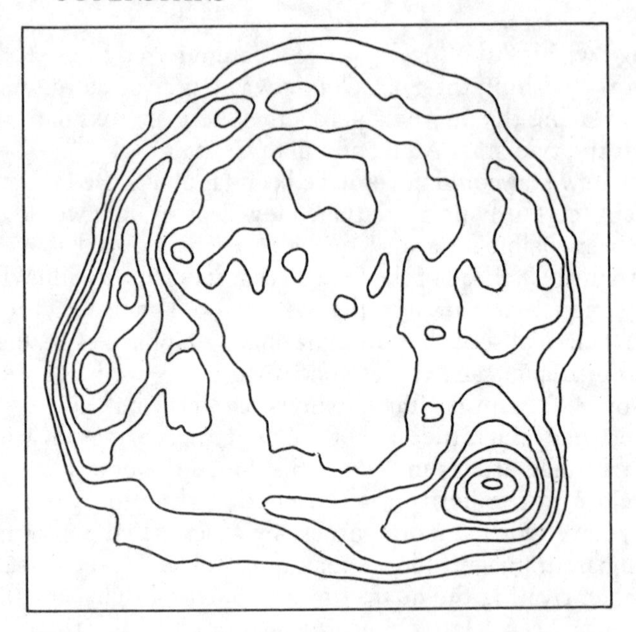

Figure 20. A radio map of the supernova remnant RCW 86.

velocity material was produced in the new star of 185, then it could not have been a supernova—at least of the form we have come to understand. Planetary nebulae are believed to be produced by unstable red giants or by normal novae; what process could produce fast filaments emanating from a planetary nebula remains a mystery. If the 185 new star had been a nova, its extreme brilliance would have placed it very nearby indeed—and a 20-month duration to decrease from its maximum magnitude to the naked-eye limit would be quite unknown for a nova. The nova/planetary nebula/ fast filament link to the 185 new star really does seem rather unlikely, especially since we do have an acceptable genuine remnant of about the required age, in the area of sky of interest—even if its position does not fit as exactly the historical description of the 185 new star lying "within" Nan-mên as does the fascinating planetary nebula. RCW 86 must remain the *probable* remnant of the supernova of 185.

The record of the 393 new star is brief, with no mention of brightness or color: "18th year of the t'ai-yüen reign-period, 2nd month, there was a guest star lying within the asterism Wei, until the 9th month when it was extinguished." The months of appearance and disappearance correspond to February/March and October/November of that year—a duration of about seven to eight months, making it a highly likely supernova candidate. The astrological prognostication that followed is interesting: "The interpretation (of the appearance of the new star) was military activities and deaths at Yen. In the 20th year Mu-jung Ch'ui and Hsi-pao attacked Wei, but were defeated and more than 10,000 men were killed. In the 21st year Mu-jung Ch'ui died and eventually his state was ruined."

The Chin dynasty, during which the new star blazed forth, was extremely weak and torn by political dissent. As early as 317 the western half of the empire was lost to barbarians. Hsiao-wu, who was emperor at the time, reigned from 373 to 397. He was by all accounts an incompetent ruler who did nothing to stem the growing tide of rebellion. When just 34 years of age, he was murdered, while hopelessly drunk, by his favorite concubine. The dynasty dragged on for another 23 years of rising crisis before its overthrow. Despite the political turmoil of the times, the astronomical records again show remarkable reliability when judged by the accuracy of the planetary data.

The asterism Wei within which the new star lay, the Tail of the Dragon in Oriental astronomy (and the tail of the Scorpion in Occidental astronomy), is well defined (see Figure 21). Unfortunately, it embraces a large area of sky, and several supernova remnants lie within its periphery. Thus, an unambiguous identification of the 393 new star with a particular remnant is not possible. It is a pity we have just a single brief record for the 393 new star. But the next historical superstar was to leave an impressive written legacy.

The new star of 1006 was the most extensively recorded from the whole period of written history before the Renaissance. Numerous surviving records from the Far East, the Arab lands, and Europe describe the star's extreme brilliance and duration in terms that leave little doubt that it must indeed

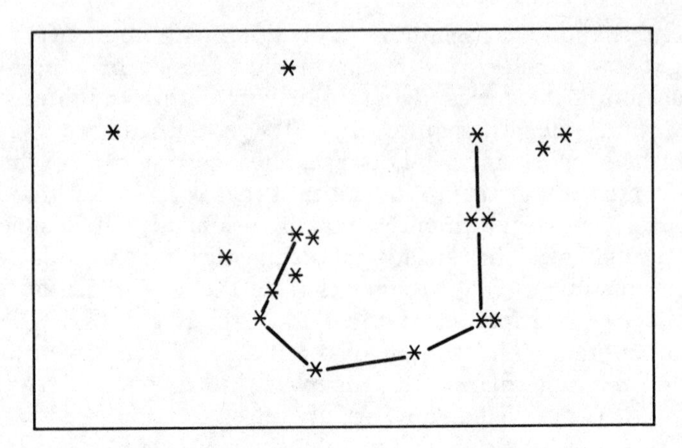

Figure 21. The asterism Wei—the "Tail of the Dragon."

have been the most spectacular stellar outburst ever seen—at any rate, as far as written records are concerned.

The European records are interesting, although only one, by virtue of the geographical location of the sighting, provides information of astronomical value. This is an account in the annals of the Benedictine monastery of St. Gallen in Switzerland. Like so many other monastic records, the St. Gallen chronicles were mainly concerned with local affairs, and there is no reason to suppose that the sighting of the new star was made anywhere other than in the immediate locality of St. Gallen. At this period, before the introduction and widespread use of paper in Europe, the histories were written on pieces of vellum. Because this was scarce and expensive, the pieces were often scraped clean so that they could be used again, and there is evidence of this on many ancient documents. It is not surprising, therefore, that only the most spectacular events from any particular year found their way into the final monastic histories. Such was the classification of the 1006 new star, which was rated by the St. Gallen chroniclers as comparable with a severe famine from the previous year and a serious plague in the year that followed:

A.D. 1005: Behold there was a famine, which during any century was no more severe.

A.D. 1006: A new star of unusual size appeared, glittering in aspect, and dazzling the eyes, causing alarm. In a wonderful manner this was sometimes contracted, sometimes diffused, and moreover sometimes extinguished. It was seen for three months in the inmost extremities of the south, beyond all the constellations which are seen in the sky.

A.D. 1007: There was a serious plague, which with sudden death devastated the people far and wide.

To the south of St. Gallen lie the Alpstein mountains dominated by the 8,215-foot summit of the Säntis, the most famous peak in that region. If indeed the star could be seen from St. Gallen, the horizon defined by the mountain range provides a valuable southern limit to the possible location of the new star.

And since the record refers specifically to the star being "in the inmost extremities of the south," another limit just a few degrees (say, five) north of the southern horizon might be set. Nothing else of positional value can be obtained from the European records, although a few interesting points are raised. For example, in the annals of the monastery of St. Sophie, Benevento, we find this entry: "A.D. 1006—In the 25th year of our master Pandolphus, and the 19th year of our master Landolphus, his son, a very brilliant star shone forth, and there was a great drought for three months." Again the annals are mainly concerned with local events, and there is every reason to believe that the star was seen in Benevento.

The following three records may have been inspired by sightings of the new star, although all contain inconsistencies, often of date, but also of position. The common reference to the object as a comet merely indicates that the observers were ignorant of its stellar nature. In the writings of Alpertus of Mertz, France, we read:

> Concerning the appearance of a comet, and famine, and mortality: Three years after the king was raised to the throne of the kingdom, a comet was seen in the southern part of the sky with a horrible appearance, emitting flames this way and that. In the following year, a most terrible famine and mor-

> tality took place over the whole Earth, with the result that in many places on account of the multitude of the dead and the weariness of those who were burying them, the living, still dragging their breath and struggling with what strength they had, were overwhelmed along with the dead.

Alpertus was a chronicler. The king referred to was Henry II, who ascended the throne in 1002 following the death of Otto III—the error of one year in the account is probably unimportant. The direction would certainly fit the 1006 new star, and Alpertus may represent a third original source from Europe.

An early suggestion that a new star may merely be the brightening of an old star is found in the history of Raoul Glaber:

> During the reign of this king, there appeared in the sky at the west one of those stars which are called comets; the phenomenon commenced in the month of September, one evening at nightfall, and lasted for nearly three months. Shining with a very vivid brightness, it occupied a great part of the sky with its light and set towards cockcrow. As for knowing whether it was a new star which God has sent, or a star whose light he had simply multiplied many times in order to signify a portent, only the One who in his wisdom rules all things best could say. What is always certain is that each time we see produced a prodigy of this sort, a little afterwards follows visibly on these something astonishing and terrible.

The exact date referred to is somewhat uncertain, although it seems to be about 1006. The initial direction (west) is inconsistent with other accounts of the 1006 star, as are the month of appearance and the duration throughout the night. Despite these disparities, it is possible that Glaber's account refers to the event but was written so long afterward that it produced a confused story as details were forgotten. What is particularly interesting is his interpretation of the phenomenon as either a new star or one multiplied in brightness, both concepts being contrary to the doctrine of celestial perfection encouraged by the Church at that time. As Glaber himself recounts in his history, however, he was anything but an orthodox cleric, continually having to move from one monastery to another. It was to be several centuries before the world would share his enlightened attitude. Perhaps Raoul Glaber deserves a place in books on the history of astronomy as possibly the first to suggest that a new star is really an old one multiplied many times in brightness.

A particularly picturesque account of a vision seen by a certain monk, Ademar of Chabannes, might possibly have been inspired by a sighting of the 1006 new star. As it stands, the record is beyond astronomical interpretation. The exact date is uncertain, since the account was written many years later, but was probably close to 1006. The vision, like the new star, was seen in the south. The record reads:

> In these times there appeared signs in the stars, harmful droughts, excessive rain, excessive plagues, and grievous famines, and many eclipses of the Sun and Moon, and the river Vienne for three nights dried up at Limoges for two miles. And the above-mentioned monk Ademar, who at the time was living with his esteemed uncle Ademar at Limoges in the monastery of St. Martial, was awakened during the night; while he was looking at the stars outside, he saw in the south in the height of the heaven a large crucifix as if fixed in the sky, and the figure of the Lord hanging on the cross weeping with a large river of tears. But he who saw these things was astonished, and could do nothing but pour forth tears from his eyes. He saw the cross itself and also the figure of the Crucified, with a fiery and very blood-like colour throughout half an hour of the night until it closed itself away in the sky. And what he saw he always hid in his heart until he wrote it down here, and the Lord is his witness that he saw it.

If Ademar indeed witnessed the 1006 new star, perhaps some other well-documented visions in history might also have been inspired by the sighting of astronomical phenomena.

The brilliance of the new star is given particular emphasis in records from the Middle East. A Syrian source says, "There appeared a great star resembling Aphrodite in greatness and splendor in the zodiacal sign Scorpio; its rays revolved and gave out light like that of the Moon." In a document from Iraq we read: "Among the incidents in that year a large star similar to

Venus in size and brightness glittered to the left of the *qibla* [the local direction of Mecca]. Its rays on the Earth were like the rays of the Moon." Without doubt, however, the most useful and the most interesting record from the Middle East is that of the scholar Ali ibn Ridwan of Cairo (died 1061). In editing Ptolemy's astrological treatise *Tetrabiblos,* he added the following autobiographical commentary:

> I will now describe a spectacle which I saw at the beginning of my studies. This spectacle appeared in the zodiacal sign Scorpio, in opposition to the Sun. The Sun on that day was 15 degrees in Taurus and the spectacle in the 15th degree of Scorpio. This spectacle was a large circular body, 2½ to 3 times as large as Venus. The sky was shining because of its light. The intensity of its light was a little more than a quarter of that of moonlight. It remained where it was and moved daily with its zodiacal sign until the Sun was in sextile with it in Virgo, when it disappeared at once. All I have mentioned is my own personal experience, and other scholars from time to time have followed it and come to a similar conclusion. . . . Because the zodiacal sign Scorpio is a bad omen for the Islamic religion, they bitterly fought each other in great wars and many of their great countries were destroyed. Also many incidents happened to the king of the two holy cities [Mecca and Medina]. Drought, increases of prices and famine occurred, and countless thousands died by the sword as well as from pestilence. At the time when the spectacle appeared calamity and destruction occurred which lasted for many years afterwards. . . .

The association of the new star's appearance with war, drought, famine, and pestilence, both in Ali ibn Ridwan's account and in several of the European records, emphasizes the impact such stellar outbursts could have on people's minds. On the other hand, not all astrological prognostications or associations were as ominous as Ali ibn Ridwan's. Indeed, in China the chief officer of the Spring Agency in the Astronomical Bureau at the time of the supernova turned its appearance to political advantage. From the biography of Chou K'o-ming (954–1017), in the history of the Sung dynasty, we read:

> [1006] . . . a large star appeared at the west of Ti [lunar mansion]. No one could determine its significance. Some said it was a "baleful star" which portended warfare and ill-fortune. At that time Chou K'o-ming was away on a mission to Ling-nan. On his return he urgently requested permission to reply to these suggestions. He said, "I have checked [various astrological treatises]. The interpretation is that the star should be called a 'chou-po' star [an 'Earl of Chou' star], which is yellow in color and resplendent in its light. The country where it is visible will prosper greatly, for it is an auspicious star. On my way back I heard that people inside and outside the court were quite disturbed about it. I humbly suggest that the civil and military officials be permitted to celebrate in order to set the Empire's mind at rest." The Emperor approved and acceded to his request. He then promoted him to the post of Librarian and Escort to the Crown Prince . . .

One can only admire the political dexterity with which Chou K'o-ming was able to change a portent of warfare and ill-fortune to one of prosperity, but it is fortunate that he did not delve too deeply into his learned books. If he had, he might have been confused by the following conflicting definitions of a *chou-po* star in the history of the Sui dynasty: "It has a brilliant yellow color, and brings prosperity to the state over which it appears"; or, "It is large and is of a brilliant yellow color. Over the state it appears it presages military action, death, and country-wide famine so that the population must seek refuge from their homes."

Chou K'o-ming was not the only one to make political capital out of the new star's appearance. The biography of Chang Chih-po, a court official of the Sung dynasty, records:

> When the chou-po star appeared, the star-clerk reported that it was an auspicious omen, and all the officials bowed to congratulate the emperor. Chang Chih-po gave his opinion, saying that the ruler should consolidate his virtues in response to the signs seen in the heavens, so that the star, which was not bound by any law, would remain visible. He then explained the necessity of maintaining a good govern-

ment. The emperor told officials that Chang Chih-po had the welfare of the Empire at heart.

As for all the other pre-Renaissance supernovae, the Far Eastern records of the 1006 event are the lodestar enabling the position, brightness, and duration to be estimated with some certainty. An example may be found in the record from the astronomical treatise of the history of the Sung dynasty in China:

> [May 6, 1006] . . . A chou-po star was seen. It appeared to the south of Ti [lunar mansion] and one degree west of Ch'i-kuan [asterism]. Its form was like the half Moon, with pointed rays shining so brightly that one could see things clearly. It passed through the east of K'u-lou [asterism], and in the 8th month [August 27 to September 24], following the wheel of the heaven it entered the horizon. In the 11th month [November 24 to December 22] it was again seen at Ti. From this time onward it regularly appeared [between 7 and 9 A.M.] during the 11th month at the east, and during the 8th month at the south-west it entered the horizon.

The part of the sky referred to is illustrated (as in 1006) in Figure 22, which shows the asterism Ch'i-kuan (containing the bright stars Alpha, Beta, Gamma, Delta Lupus, plus some fainter stars) in Ti lunar mansion, as well as the neighboring asterism K'u-lou (Theta Centaurus plus some fainter stars), and the lunar mansion K'ang. Reference is also made to these in another Oriental record, the *Ch'ing-li-kuo,* a book presented to the Chinese emperor in 1044 and dealing with court matters during the first 80 years of the Sung dynasty. The book has this to say:

> [May 1, 1006] The Director of the Astronomical Bureau reported that at the first watch of the night, on [May 1, 1006], a large star yellow in color appeared to the east of K'u-lou at the west of Ch'i-kuan. Its brightness had gradually increased. It was found in the 3rd degree east of Ti. . . . The star later increased in brightness. According to the star manuals there are four types of auspicious stars. One of these is called a chou-po; it is yellow and resplendent and forebodes great prosperity to the state over which it appears.

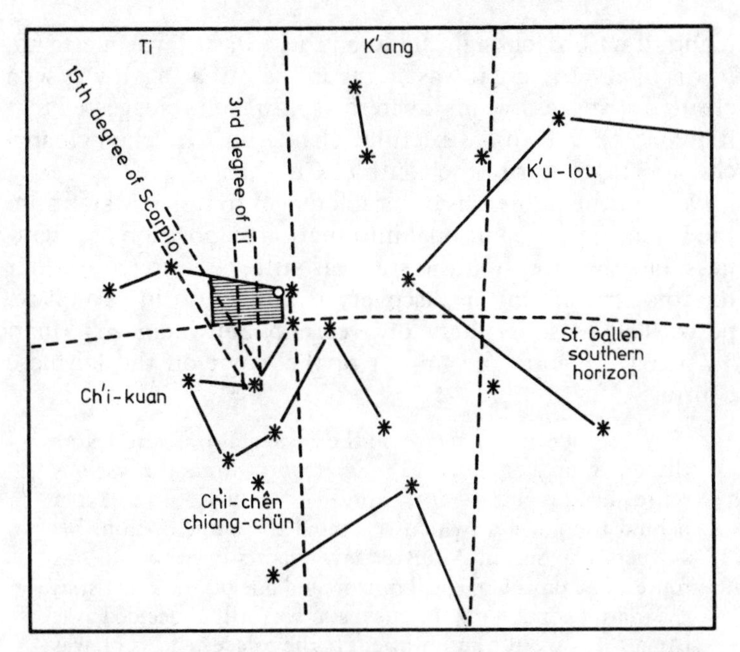

Figure 22. The area of the sky in which the supernova of 1006 was observed, with the preferred search area for its remnant shaded. The circle on the periphery of the shaded area shows the position of the remnant PKS 1459-41.

One interesting Chinese record suggests that the *chou-po* star reappeared a decade later in 1016. If we had any evidence of postoutbursts from extragalactic supernovae, we might have some confidence in the reality of this historical record. Without such evidence, one is left with the likelihood that the 1016 new star was a historically misplaced reference to the 1006 supernova.

A record from Japan placed the new star within the asterism Ch'i-kuan. This record is in the *Meigetsuki* ("Diary of the Full Moon"), the personal diary of Fujiwara Sadaie, a thirteenth-century poet-courtier. Fujiwara seems to have had a special interest in new stars, and his references to them are often particularly detailed, showing that he obviously had access to original sources, even though he does not give them. Concerning the 1006 new star, his entry reads: "[May 1, 1006] After nightfall within [chung] Ch'i-kuan there was a large guest star. It was like Mars, and it was bright and scintillating. It was seen clearly for successive nights in the south. Some suggested that it might be due to a structural change in Ch'i-chên-chiang-chün [a single star to the south of Ch'i-kuan]."

None of the numerous other references to the new star from the Far East gives additional information on position or brightness, but the attention the star was still attracting a full four months after its initial discovery is alluded to in several reports. The personal diary of a certain Fujiwarano Yukinari (971–1027) reports the impact of the event on the Japanese court:

> [July 21, 1006] . . . a report on the guest star was read before the emperor . . . [July 22] . . . reports from the various routes on the guest star were read . . . [July 30] . . . from behind the curtain [which separated the emperor from his subjects] the Second Minister gave reports from the various routes, one on the magical mirror, and one on the guest star; the matter concerning the guest star was still undecided . . . [August 8] . . . [the author] went to the palace; a decision was made on the guest star.

Fujiwarano does not tell us what the decision was, but fortunately the decision is revealed in another historical treatise:

> [August 9, 1006] a decision was made regarding the reports from the various routes on the large guest star . . . [August 15] . . . the cabinet ordered that divination be made regarding the large star . . . [September 3] . . . offerings were made to twenty-one shrines . . . [September 21] . . . the Second General came and asked for an amnesty because of the appearance of the large star.

With such a wealth of independent observations, it should be a simple task to locate accurately the position of the new star in the sky, and this proves to be the case. As already mentioned, the St. Gallen sighting provides a valuable southern limit. This is indicated on Figure 22 as is Ali ibn Ridwan's estimate that the star lay within the fifteenth degree of Scorpio. The third degree of Ti, specified in the *Ch'ing-li-kuo,* intersects the fifteenth degree of Scorpio within the asterism Ch'i-kuan mentioned in several of the other Far Eastern records. Indeed, it was argued earlier that the word "within" (chung), specified in the *Meigetsuki,* meant that the star lay entirely within the boundary of an asterism. Such an exact interpretation would not be in conflict with the Sung history record stating that the star appeared one degree west of Ch'i-kuan, since this could have meant one degree west of center (but still within the periphery of the asterism). Allowing an error of possibly up to two degrees in the estimates of those records giving specific positional details, the area of sky compatible with all the historical descriptions (except one) turns out to be quite small—it is shown shaded in Figure 22.

But what about the sole historical record incompatible with the others? This is the reference in the *Meigetsuki* that some thought the new star "might be due to a structural change in Ch'i-chên-chiang-chün"—the single star (Kappa Lupus) to the south of Ch'i-kuan. The sources of Fujiwara's information here seems to be a conversation reported in the diary of Fujiwara Kanazane covering the period from 1164 to 1200:

> [May 8, 1169] Clear sky. Abe Yasuchika, the Assistant Astronomer, came in the evening. . . . The following was said in conversation: During the time of emperor Dai Sanjō In [reign 1011–1016] there was an argument between Ariyuki and Morohira. Ariyuki said that what appeared within the neighborhood of Ch'i-chên-chiang-chün was a guest star, but Morohira said that it was not a guest star. Rather he attributed this to the complete appearance of all the stars in the asterism Ch'i-chên-chiang-chün. Morohira found more support in their discussion and Ariyuki was much enraged. It was prognosticated that an important national event would take place within the next three days, and during that time the emperor Sanjō In had cause for anxiety. Hence what Morohira said was regarded as the correct explanation . . .

Just how accurate an account Abe Yasuchika would have of an argument which took place some 150 years before his own time is impossible to say. What is certain is that the asterism Ch'i-chên-chiang-chün (the "Cavalry General") is only a single star, yet Morohiro's case was based on the appearance of all the stars in the asterism. Is it possible that Ch'i-kuan was intended? Certainly the 1006 event was a guest star, so that if this was indeed the subject of the discussion, then Ariyuki had every reason to feel enraged. The whole record of the argument appears confused, and the positional discrepancy with the other historical descriptions is probably unimportant.

Having estimated the position of the new star to within a few degrees, it is of interest to see whether the historical records allow the brightness and duration of the event to be determined. The duration of the event must have been several years. The fact that many records specify a duration of three to four months is misleading; this was merely the period until the star "set," after which it was for about three months a daytime object. When next it rose in the night sky its brightness would have diminished to the point where its reappearance might escape the attention of all but the most assiduous stargazer. The record from the astronomical treatise of the history of the Sung dynasty refers to regular setting and rising, implying that in China at least the gradually fading new star was tracked for several (that is, greater than two) years, until it reached the limit of naked-eye visibility.

If it took several years to fade from maximum brightness, how long did it take to reach maximum brightness after its initial discovery on or about May 1, 1006? This is impossible to determine with certainty, although the *Ch'ing-li-kuo* record suggests that the period of brightening may have been extensive. The report from the director of the Astronomical Bureau dated May 30 suggested that the new star's brightness had gradually increased (since discovery May 1). Then the star "later increased in brightness." Could *later* mean the star brightened even after the day of the report (May 30)? If so, the new star was discovered more than 30 days before reaching maximum brightness.

Finally, we come to the question of how bright the new star was at maximum. The numerous records of the event, and their extravagant allusions to its spectacular nature (for example, "glittering in aspect and dazzling the eyes"), bear witness to its extreme brilliance. The comparisons with the Moon allow a more specific estimate of brightness to be made: "it gave out light like that of the Moon"; "its rays on the Earth were like the rays of the Moon"; "the intensity of its light was a little more than a quarter of that of moonlight"; "its form was like the half Moon, with pointed rays shining so brightly that one could see things clearly." (The quarter-illuminated Moon is of magnitude −8.5. One can start to discern distant objects by the light of the Moon when the age is about five days with magnitude close to −9, the half Moon being nearly −10 and the full Moon −12.5.) The comparisons with the Moon would therefore seem to suggest a brightness exceeding −8.5. The new star of 1006 never attained an altitude (above the southern horizon) of more than 17.5 degrees at K'ai-feng, the Chinese capital of the time; allowing nearly a magnitude for atmospheric extinction, an apparent magnitude at maximum of −9.5 would seem to be a conservative estimate. This would make the 1006 new star by far the brightest witnessed during the period of recorded history. Its distance from the Earth was probably no more than some 3,000 to 4,000 light-years.

Extreme brightness and long period of visibility make it certain that the 1006 new star was a supernova. This being so, and with the position being fairly well determined from the historical descriptions, an unambiguous identification of the supernova remnant ought to be possible. This is indeed the case. On Figure 22, the circle on the northwestern periphery of the shaded search area indicates the position of a radio object (called PKS 1459-41) that shows the characteristic ring structure, radio spectrum, and polarized emission of a supernova remnant. There is no other supernova remnant within the shaded area of Figure 22. PKS 1459-41 was discovered with the giant 210-foot-diameter radio telescope at Parkes in 1965. A radio map of the object is shown as Figure 23. The association with PKS 1459-41 fits the St. Gallen sighting particularly well, and the apparent path of a star at the location of the remnant is shown in Figure 24. The remnant of the brilliant 1006 supernova has been discovered in X rays, and the X-ray map obtained by the *Einstein* spacecraft shows excellent posi-

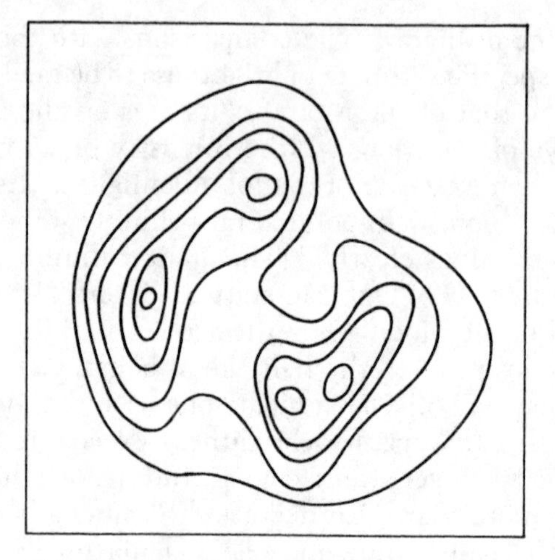

Figure 23. A radio map of the remnant of the brilliant supernova of 1006.

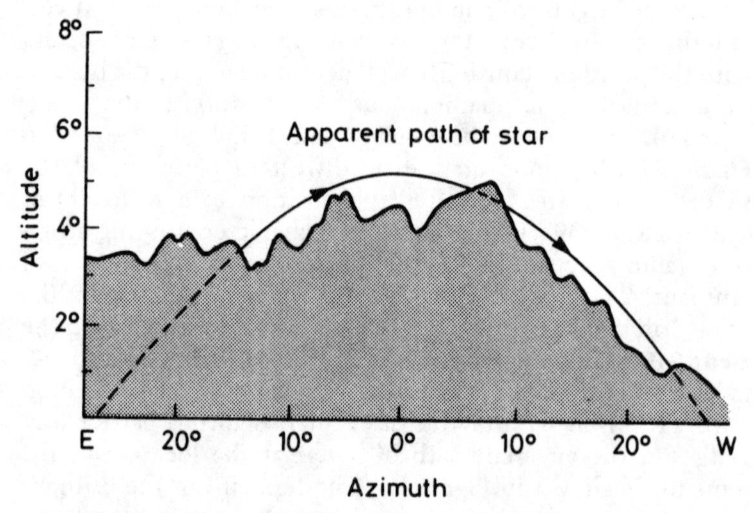

Figure 24. The apparent path of a star at the location of the supernova remnant PKS 1459-41 in relation to the St. Gallen horizon in 1006.

tional agreement between regions of bright radio and X-ray emission. Faint, delicate optical filaments have been found along the northwestern edge of the remnant in the gap in the radio/X-ray shell—there is no optical nebulosity coincident with the bright radio/X-ray regions. There can be no doubt that we are witnessing the expanding remnant from the site of the gigantic stellar explosion that made such a remarkable and well-documented impact on the troubled world of 1006. Such is the legacy of the superstars.

7

A SUPERSTAR IS BORN

There is nothing born but has to die.

Thomas Carlyle

IBN BUTLĀN was a Christian physician, originally of Baghdad and later of Cairo, and for a brief period (including the year 1054) of Constantinople. It is probable that his sojourn in Constantinople was brought about by his flight from the devastation of famine and pestilence. Like many other physicians of the period, Ibn Butlān was obsessed by the possibility that terrestrial catastrophes, like famine and plague, were connected with cosmic events such as the appearance of comets. Not surprisingly, therefore, a report he wrote covering the period from April 1054 to April 1055 makes mention of a spectacular new star of that year:

> One of the well-known epidemics of our own time is that which occurred when the spectacular star appeared in Gemini in the year [April 1054 to April 1055]. In the autumn of that year fourteen thousand people were buried in the cemetery of the church of St. Luke, after all the other cemeteries in Constantinople had been filled. Then, in midsummer the Nile was low and most people in Old Cairo and all the strang-

ers died, except those whom Allāh willed to live. The epidemic spread to Iraq and affected most of the population, and the land was laid waste in the wake of contending troops, and this continued until the year [1062]. In most countries people fell ill with black-bile ulcers and swelling of the spleen. The usual arrangement of the rise and fall of the fevers was altered, and the order of the crises was upset, so that the rules of the science of prognosis had to be changed. As this spectacular star appeared in the sign of Gemini which is the ascendant of Egypt, it caused the epidemic to break out in Old Cairo when the Nile was low, at the time of its appearance. Thus Ptolemy's prediction came true: "Woe to the people of Egypt when one of the comets appears threateningly in Gemini!" Then when Saturn descended into the sign of Cancer, the destruction of Iraq, Mosul and Jazīra was completed; Diyār Bakr, Rabīca, Mudar, Fārs and Kirmān, the lands of the Maghrib, Yemen, Fustāt and Damascus/Syria were upset; the affairs of the kings of the world were disturbed; and wars, famine and epidemics abounded. And this confirmed the wisdom of Ptolemy in saying: "When Saturn and Mars are in conjunction in the sign of Cancer, the world will be shaken."

There is little of astronomical importance in Ibn Butlān's report. It tells us nothing about the precise location of the star, its exact date of discovery, its brightness, color or duration. The rediscovery of the report, however, proved of tremendous interest to astronomers, since it represented the only positively identified eyewitness account from outside the Far East of what we now know to have been a supernova in the year 1054. (In making this statement, it is recognized that the American Indian rock-art depictions may eventually be related positively to the 1054 new star.)

There is one possible allusion to the new star from Europe. In a chronicle compiled in Bologna about 1476, but based on earlier sources now presumably lost, the following passage occurs: "In the year [1058] Henry the third had reigned nineteen years. He first came to Rome in the month of May. At this time there was a famine and death throughout the whole land. . . . At this time a most clear star appeared in the first circuit of the

Moon. . . ." What the first circuit of the Moon could mean—perhaps positional proximity to the Moon, which was certainly the case for the 1054 new star—remains uncertain. The year 1058 is impossible—Henry III died in 1056. Could the "most clear star" have been the supernova of 1054, its date transcribed in error by the chronicler? Even if this was the case, is it surprising that the Europeans failed to provide other written records? Not really. The new star of 1054 was very much fainter than the brilliant 1006 supernova, and even for that spectacular event there were few European records. Interest in astronomy was limited at this time, and scholars were ignorant of the form of the heavens. The star would have been seen in the early-morning sky, close to the ecliptic, and might merely have been mistaken by some for Venus, which is seen as a morning (or evening) star. Even the most avid star watchers in Europe at this time were confused by sightings of Venus. A canon of the Cathedral of Vysehrad in Prague recorded a host of astronomical sightings for the interval from 1126 to 1142; in 1131 he reported two strange stars, noting, "I do not believe that there are men who know which among them was called Lucifer [Venus]." But from his descriptions and positions it is probable that both sightings were of Venus. If even the keenest of medieval European astronomers/scholars had such a poor appreciation of the heavens, it is really not surprising that the 1054 new star went unheralded. Once again, however, we find an impressive list of observations from the Orient.

A particularly fascinating record from China is contained in a volume known as ("The Essentials of Sung History") *Sung-hui-yao* compiled sometime during the Sung dynasty. It records how the astrologer Yang Wei-tê first reported the new star's appearance to the emperor, on August 27, 1054, almost eight weeks after the star's discovery:

> Yang Wei-tê said, "I humbly observed that a guest star has appeared; above the star in question there is a faint glow, yellow in color. If one carefully examines the prognostications concerning the emperor, the interpretation is as follows: The fact that the guest star does not trespass against Pi [an asterism] and its brightness is full means that there is a

> person of great worth. I beg that this be handed over to the Bureau of Historiography." All the officials presented their congratulations and the Emperor ordered that it be sent to the Bureau of Historiography.

The record suggests that Yang Wei-tê may have been using the appearance of the guest star to further his own political aspirations, since there would seem to be no other obvious reason for his referring to the asterism Pi, more than 15 degrees from where we now know the new star appeared. Furthermore, his allusion to the star as yellow is probably unreliable: yellow just happened to be the imperial color of the Sung dynasty, which may also explain why many of the Chinese records of the 1006 supernova referred to it as yellow too.

The date of first appearance of the new star is given in the astronomical treatise of the history of the Sung dynasty: "[July 4, 1054] A guest star appeared approximately several inches [that is, several tenths of a degree] to the southeast of the asterism T'ien-kuan. After a year and more it gradually vanished." The month of first appearance is confirmed elsewhere in the history: "[April 17, 1056] The director of the Astronomical Bureau reported that since the month [July 1054] a guest star had appeared in the morning at the east, guarding T'ien-kuan, and now it has vanished." The implication is that the star may only just have vanished at the time of the report, so that its total duration would then have been about 21 months, which fits the astronomical treatise report of it vanishing after a year or more. The month of disappearance is also given in another section of the *Sung-hui-yao,* which provides the additional information that it was a daytime object at discovery and suggests that it was seen for 23 days in daylight: "[July 1054] The guest star appeared in the morning in the east guarding T'ien-kuan. It was visible in the daytime, like Venus. It had pointed rays on all sides, and its color was reddish-white. Altogether it was visible [in daylight] for 23 days." It is interesting to see how the colorimetry of this report differs from that of Yang Wei'tê.

From these records we may conclude that the new star was first sighted on July 4, 1054. At this time it was an early-morning object rising just ahead of the Sun, and it would have passed almost directly overhead at about ten o'clock in the morning. It was visible in daylight for a total of 23 days, and after as many months the star had faded to below the limit of naked-eye detection. A duration of 21 months means that the star must have disappeared and reappeared after conjunction with the Sun, although no record refers specifically to this.

All the records quoted above emphasize the proximity of the new star to the asterism T'ien-kuan. T'ien-kuan (the "Celestial Gate") was an unusual asterism in that it contained just a single star. Ancient accounts were definite about this, and this statement is typical: "T'ien-kuan is a single star. . . . It lies where the Sun and Moon move." This means that T'ien-kuan lies on (or close to) the ecliptic. T'ien-kuan can in fact be identified, from both historical records and ancient star charts, as the star we now know as Zeta Tauri—an isolated third-magnitude star, there being no star brighter than fifth magnitude within eight degrees (see Figure 25).

There is a single historical record of the new star with a positional description inconsistent with those above. This is contained in the *Memoirs of the Liao Kingdom,* a seminomadic kingdom in the extreme north of China from 937 to 1125. The record reads: "[August 28, 1055] The King died. . . . Previously there had been an eclipse of the Sun at midday, and a guest star appeared at the asterism Mao. The deputy officer of the Bureau of Historiography, Liu I-shou, said, 'This is an omen that King Hsin-tsung will die.' The prediction indeed came true."

The solar eclipse referred to must have been that of May 10, 1054. This was total in central China and would be fairly large in the north, and there was no other large eclipse until 1061. The date of the guest star clearly agrees with that observed elsewhere on July 4, 1054. Yet the asterism referred to in the Liao account, Mao (the Pleiades), is some 20 degrees from T'ien-kuan. As the record is concerned mainly with the passing of the king, the writer may not have been concerned with astronomical precision. The Pleiades is a familiar star group, but perhaps he was unable to identify nearer asterisms, or per-

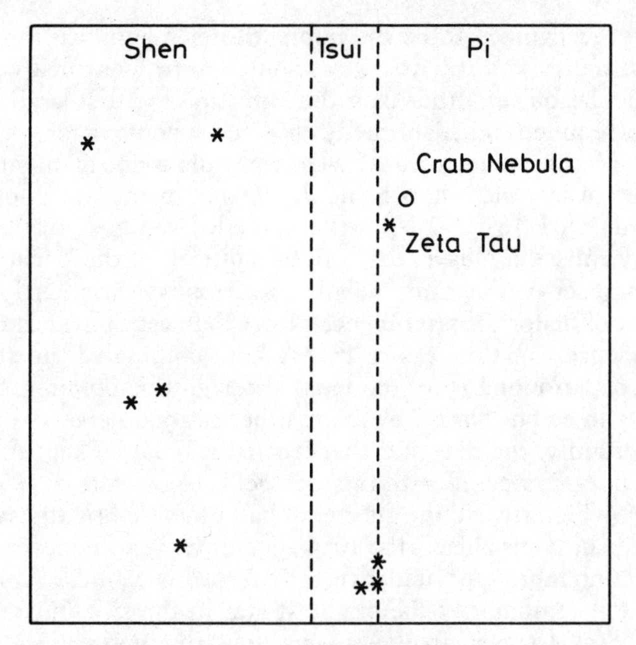

Figure 25. Stars in the vicinity of Zeta Tau.

haps the reference to Mao was purely astrological. Whatever the reason, the discrepancy is unimportant.

The possible brightness of the new star at maximum can be inferred by its discovery in daylight and its comparison with Venus. The magnitude of Venus varies from about −4.5 to −3.5. Far Eastern sightings of the planet in daylight are fairly frequent but are still rare enough to be regarded as omens. The apparent ease of the new star's detection in daylight suggests that it must have been at least of magnitude −4.

The impact that a number of crude observations made nearly a thousand years ago has had on modern astronomical research seems scarcely believable. The reason for this is that the Crab Nebula, the plerionic remnant of the new star, has become one of the most extensively studied objects in astronomy. No other object has stimulated so many new theories or so readily provided the observational means to test them. It was the first object recognized to be the remnant of a supernova; the first radio source and the first X-ray source to be identified with a particular object (other than the Sun); and the first (and so far only) remnant of a historically observed supernova for which a pulsar has been detected. The Crab Nebula is one of only three supernovae—the others being the Vela remnant and an object called MSH 15-52—displaying extended radio, optical, and X-ray emission plus a pulsar. But despite all of these features, the impact of the Crab Nebula in astrophysics is partly the result of historical prominence rather than astrophysical transcendence. For this reason, it may have dominated the attention of astronomers for too long, although this opinion is unlikely to be one shared by many other astronomers.

Certainly, the new star that created it must be kept in perspective. Compared with the spectacular new stars of 185 and 1006, which rivaled the quarter to half Moon in brightness and set the heavens ablaze, the 1054 supernova was a nonevent, an object no more spectacular than Venus. This would, of course, make it temporarily the brightest star in the sky, but not so bright as to cast shadows, spectacularly increase the sky brightness, or necessarily cause the alarm its brilliant predecessor did. Yet for some unexplainable reason, authors of many astronomy books have been tempted to exaggerate grossly the spectacle of its appearance and have found it necessary to use the 1054 outburst as an apparent prototype historical supernova. Perhaps the reason is sheer ignorance of its unimpressive nature compared with the new star that preceded it by just 48 years.

Optically, the Crab Nebula's tangled web of filamentary structure must be one of the most familiar of astronomical objects. The discovery of the nebulosity has been attributed to an English amateur astronomer and physician, John Bevis. Bevis, working from a small observatory at his own home, slaved for many years at preparing an atlas of stars. In 1731 he found and recorded the strange nebula in Taurus. Unfortunately, the atlas was not finally printed until 1786, after Bevis's death. By this time the nebula had been "rediscovered" and reported by the French comet hunter Charles Messier. Messier had been searching for Halley's comet, predicted to reappear in 1758. He reported his supposed rediscovery of the comet in an entry for the night of August 28, 1758: "I found the comet of 1758, which ought to be between the horns of Taurus, below the southern horn, a small distance away from the star Zeta of this constellation. It appeared as a whitish elongated light spot, resembling a candle in its shape, and containing no stars." But Messier was wrong. Halley's comet was not sighted in that year until Christmas day, and then by a peasant without the aid of a telescope. Messier resolved not to be so misled again and set about listing all objects that might cause confusion in future cometary studies. His initial list was published in 1771, as the first catalog of nebulae, with the nebula in Taurus, subsequently known as M1, listed first.

M1 seems to have become the Crab Nebula about a century later, and the name may have been derived from the fact that Lord Rosse's drawing of the object in 1844 made it look rather like a crab. At any rate, the name has certainly been in general use from about that time. The Crab Nebula was first successfully photographed with a 20-inch reflector telescope in 1892, and it has become probably the most commonly photographed astronomical object. By 1920, high-quality photographic plates of the Crab had been obtained often enough to make it clear that on plates separated sufficiently in time there had been a discernible outward movement of the prominent filaments of the nebula. By reversing these motions, it could be estimated that the expansion of the Crab Nebula had begun 800 to 900 years earlier. These observations were made by John Duncan of the Mount Wilson Observatory, who published a short note entitled "Changes Observed in the Crab Nebula in Taurus" in 1921. In the same year, Lundmark published his early results on bright novae recorded in Chinese histories, including the new star of 1054. Lundmark must have been well aware of the proximity of the new-star position and the Crab Nebula but didn't make the association in print. This was left, as was so much else in astronomy in the twenties, to Edwin Hubble. In 1928 Hubble made the link between Duncan's and Lundmark's papers in a short article. But in the aftermath of the great debate on the island-universe controversy, astronomers apparently could not spare the time to investigate the proposed association. More than a decade passed, until Nicholas Mayall asserted, "It may be said that the identification of the Crab Nebula as a former supernova possesses a degree of probability sufficiently high to warrant its acceptance as a reasonable working hypothesis." But new measurements by Duncan of the expansion of the nebula suggested "with considerable uncertainty, the year 1172 as the date of outburst"—not 1054!

The story now moves to Holland and to work instigated by the great Dutch astronomer Jan Oort. He persuaded a colleague at Leiden University, the Sinologist J. Duyvendak, to reinvestigate the Chinese records. In a footnote to Duyvendak's written interpretation of the records, Oort drew attention for the first time to a discrepancy in one of the records. While most records make reference to the new star "appearing in the vicinity of" or "guarding" T'ien-kuan, the single record that attempts to be precise (the astronomical treatise of the Sung history) states that the star was positioned several tenths of a degree southeast of Zeta Tauri. Yet the only known supernova remnant in the vicinity of Zeta Tauri, the Crab Nebula, is about a degree northwest of Zeta Tauri, the next closest supernova remnant to Zeta Tauri being about six degrees away. Oort wrote:

> It should be noted that a slight discrepancy exists between the position of the Crab nebula and that of the "guest star" reported in the History of the Sung Dynasty. According to this report the star was situated south-east of Zeta Tauri, whereas, if the identification is correct, it should have been situated a little over a degree to the north-west of Zeta Tauri. Prof. Duyvendak has confirmed that it is unlikely that a mistake of copying was made in the Chinese sources.

Oort sent Duyvendak's paper out of Nazi-occupied Holland, via Sweden, to Mayall in the United States. Duyvendak's paper plus a companion paper by Mayall and Oort supporting the proposed association of nebula and new star despite the directional discrepancy were published in 1942. From that date, the vast majority of astronomers accepted that the 1054 new star gave birth to the Crab Nebula.

In 1942, Walter Baade, not prepared to let the matter rest, repeated Duncan's measurement. Like Duncan, he found that a simple extrapolation backward of the expansion gave a date of origin at least a century after the new-star date. But Baade was willing to accept the proposed association of the nebula and new star: "The important new data about the nova of 1054 which Professor Duyvendak has recently made available leave hardly any doubt, as Oort and Mayall have shown, that this star is the parent of the Crab Nebula. . . ." Baade concluded that the discrepancy in the date of origin required an acceleration in the expansion. But what caused the acceleration? Baade noted that "difficulties arise, however, when we try to understand it in terms of known forces." The identification of the mystery force was to wait nearly 30 years until the discovery of the pulsar at the center of the Crab and the realization that it could provide all the energy required to have accelerated the nebula.

We are then left with the directional discrepancy in the Sung history. In view of the overwhelming evidence now available to associate the Crab Nebula with the 1054 outburst, it might be best to dismiss the discrepancy by saying that the Chinese astronomers inadvertently transposed the relative positions of the new star and the asterism. There is in fact some evidence that the quality of Chinese observational astronomy at this time was at a low ebb and that the astronomical officials poorly understood the structure of the heavens. An intriguing section of the *Dream Pool Essays,* written by the later astronomer Royal Shen Kua, reads:

> In the Huang-Yu reign-period [between 1049 and 1053] the Ministry of Rites arranged for the examination candidates to be asked to write essays on the instruments used for gaining knowledge of the heavens. But the scholars could only write confusedly about the celestial globe. However, as the examiners themselves knew nothing about the subject either, they passed them all with a high class.

It is highly likely that astronomical officials of such questionable caliber were able to produce a directional error.

A larger fraction of the new power of modern astronomy has been focused on the Crab Nebula than on any other object. It is a plerionic supernova remnant showing central brightening rather than the more usual doughnut shape. It is a strange coincidence that the next supernova to be recorded historically should also produce a plerionic remnant with some similarities to the Crab. But if the 1054 supernova was a fairly modest spectacle, the 1181 new star was positively puny, perhaps not even exceeding the zero magnitude. Again we have the diligent Orientals to thank for recording it.

All the references to the 1181 new star are brief, and all those that give useful details are quoted below. Figure 26 shows the various asterisms referred to in the records. First, in the history of the Sung dynasty, we read: [August 6, 1181] . . . a guest star appeared in K'uei-hsiu [lunar mansion] and invading [*fan*] Ch'uan-shê [asterism] until [February 6, 1182], altogether 185 days; only then was it extinguished." This rather terse record gives a duration of six months, which strongly suggests that the new star was a supernova, as well as an allusion to position.

Disaster had struck the empire in the form of the warlike Kin from the north who had set up a dynasty in 1127 after overrunning the Liao kingdom. The Sung emperor was forced to agree to the partitioning of China in 1141. In the separate history of the Kin dynasty, the new star is recorded with a discovery date five days later than in the south: "A guest star was seen in the vicinity of Hua-kai [asterism], altogether for 156 days." The Kins had apparently not only discovered the new star later than their rivals in the south but also lost sight of it three weeks earlier. Differing weather conditions might explain this, although since it would have been the dry season in the north at this time, this should not have been the reason. A more likely explanation is that the star's brightness remained near the limit detectable by the unaided eye between the last Kin observation and the last Sung observation.

The Japanese courtier Fujiwara Sadaie noted the new star in his diary entitled the *Meigetsuki,* again without quoting his source, which must have been an independent one since he had been born just one year before the new star's appearance. His entry refers to a date just one day after the new star was first sighted in southern China: ". . . a guest star appeared at the north, near Wang-liang [asterism] and guarding Ch'uan-shê." Again from Japan, the star is recorded in the history of the Kamakura Military Government: "[Between seven and nine o'clock in the evening] a guest star was seen in the northeast. It was like Saturn and its color was bluish-red and it had rays. There had been no other example since that appearing in [1006]." The history seems to have overlooked records of the 1054 supernova in making this last statement or interpreted its appearance differently. Although the direct comparison with the 1006 supernova might appear to imply exceptional brilliance, it is doubtful whether after an interval of almost two centuries such a comparison is meaningful. None of the other records makes reference to extreme brightness, nor is it likely with a duration for the event of a mere six months, compared with several years for that of 1006. The star seems to have escaped detection in Korea or the Occident, again suggesting that it was an unexceptional object.

The comparison with Saturn in the Military Government history is difficult to interpret. At the time of discovery in Japan, the only planet visible was Mars (with magnitude about −1), fairly high in the southeast. Jupiter (with magnitude about −2) would have been visible near midnight, while Saturn (magnitude about +1) would have been seen for only a short time before sunrise. The guest star was circumpolar and observable throughout the night but, as the record says, in the northeast at the hour of discovery. Perhaps it most closely resembled Saturn in brightness but certainly not in color, since the planet is usually seen as whitish. But a first-magnitude star could hardly have been expected to produce the claimed rays, protuberances that appear surrounding all stars brighter than about zero magnitude and caused by distortion of light within the eye. The question of the star's brightness must remain open, the only possible estimate being that it was probably within a magnitude or two of zero. Such a maximum brightness is consistent with the recorded duration of six months.

Now we come to the question of position. The descriptions in the various records are confused, placing the star in the vicinity of several asterisms. The component stars of these asterisms can be identified with some certainty from ancient star charts. The asterisms of interest, shown in Figure 26, will now be considered in turn. Hua-kai, referred to in the Kin record, is the "Gilded Canopy," suspended above the important asterism Ta-ti, the "Imperial Throne" (the polestar, Polaris). The shape of Hua-kai is so characteristic of a canopy (umbrella) that it can be readily recognized even though the constituent stars are quite faint, all about fourth to fifth magnitude. Hua-kai lies in the lunar mansion K'uei-hsiu, referred to in the Sung account.

Fujiwara Sadaie placed the new star near Wang-liang. This asterism, which comprises many of the bright (second to

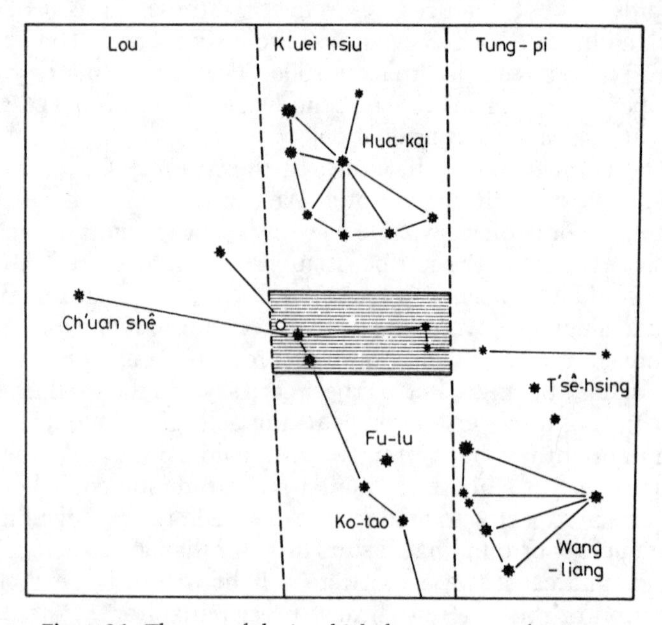

Figure 26. The area of sky in which the supernova of 1181 was observed. The preferred search area for its remnant is shaded, and the circle on its left edge shows the position of the remnant 3C 58.

fourth magnitude) stars of Cassiopeia, commemorates a famous early charioteer. The star to the west (right) is the charioteer Wang-liang, the other four stars of the asterism representing his horses, T'ien-szŭ ("Celestial Quadriga"). The isolated fourth-magnitude star to the north is the asterism T'sê-hsing ("Whip Star"), the whip used by the charioteer Wang-liang. The *Meigetsuki* account has the new star guarding Ch'uan-shê, while the Sung astronomers had it invading this same asterism. Ch'uan-shê comprises nine faint stars, nearly all lying close to the naked-eye visibility limit, along a staggered row and representing inns or guesthouses. The occurrence of two independent references to Ch'uan-shê in the accounts of the new star suggest that it was probably a well-known formation, despite its faintness. This is also apparent from the frequent allusions to the asterism in many cometary records. It must be remembered that Cassiopeia is circumpolar, so that this region of the heavens (named by the Chinese the Purple Palace or Circle of Perpetual Visibility) was probably extremely well studied and well understood. When this is kept in mind, it is rather surprising that with such a diversity of asterisms mentioned in the records of the 1181 new star, none of these records mentions two other asterisms in the vicinity. The first of these is Ko-tao, a "Flying Path Across the Mountains," consisting of one third-magnitude star plus several faint ones, and crossing Ch'uan-shê. The other is Fu-lu, a single third-magnitude star representing an "Alternative Side-route."

What is to be made of these apparently discrepant positional descriptions? Experience shows that the most precise astronomical term used in any of the descriptions is *fan* ("invade," "offend," "trespass against," etc.), a word found in the Sung history. *Fan* is very frequently used in the Far Eastern records to describe conjunctions of the Moon and planets with one another or with stars. Numerical descriptions of the term are given in various of the Chinese astronomical works, for example, ". . . *fan* . . . when a celestial body comes within [seven tenths of a degree] of another, such that their rays extend toward each other." Numerous Chinese and Korean descriptions of the Moon and planets invading one another, or stars, have been carefully scrutinized. It seems that the term *fan* was rarely used unless a separation for two astronomical bodies of less than a degree had occurred. But how could such consistent precision have been achieved? Probably with the use of a sighting tube, a long, hollow cylinder, like a telescope without lenses, held to one eye. Only when both objects under study could be simultaneously seen at a single pointing position of the tube would the term *fan* be used. Without doubt, *fan* was a very precise astronomical term. Its use does not necessarily imply motion; in the context of a new-star record, the reference to "invading" is to a star, not previously seen, suddenly appearing close to one of the permanent members of an asterism.

Placing the new star within one degree of Ch'uan-shê, and within the lunar mansion K'uei-hsiu, as also recorded in the Sung history, restricts it to the shaded area in Figure 26. Such a location is perfectly compatible with the vague Kin description of the star lying in the vicinity of Hua-kai (about four degrees away), or the *Meigetsuki* description of it lying near Wang-liang (about eight degrees away). Since these two asterisms are so prominent and easily recognizable, as well as appearing to be particularly well known, reference to the new star being in the same area of sky is hardly surprising. But what is surprising is that the asterism Ko-tao, which crosses the preferred search area, is not included in any of the records. Possibly the reason for this was astrological, or perhaps reference to it was excluded in the final editing process of compiling the official histories that were mere summaries of original records. Unfortunately, we shall never know.

The only known young supernova remnants in Cassiopeia are Cassiopeia A, the remnant of Tycho's supernova of 1572, plus an object known as 3C 58. This last object lies within the shaded preferred area of Figure 26, near the intersection of Ch'uan-shê and Ko-tao. It is a plerionic remnant, very similar in many of its radio properties to the Crab Nebula. In the optical, however, it is very much fainter, with weak, soft X-ray emission. The X-ray map from the *Einstein* Observatory shows a central point source, presumably the neutron star left from the core collapse. No other suitable supernova remnants are known in this part of the sky. Thus, despite remaining doubts as to the exact location of the new star of 1181 and the fact that certain of its properties do not appear entirely compatible with a plerionic remnant younger than the Crab Nebula, 3C 58 must be considered the *probable* remnant of the supernova of 1181.

We now return to the new star of 1408, which did not make the short list of potential supernova candidates compiled earlier since the historical records do not give a duration. The official Ming dynasty history merely states: "[October 24, 1408] . . . to the southeast of Niandao (asterism) there is a star like a lamp, yellow in color. Its light is smooth and it does not move." Since it did not move, it presumably was not a comet. The writers likening the star to a lamp suggest it must have been very bright, but because no direct comparison with nearby stars or planets was attempted, we cannot be certain. The only hint of duration comes from Chinese provincial records and the writings of a Japanese diarist. The provincial government record reads: "[September 10, 1408] . . . there is a star as large as a lamp, color bluish-white and brilliant, appearing in the east." Here we don't have even an approximate position, and the description of color differs, but presumably the record refers to the same new star. The date is some six weeks earlier than the official history. In the *Diary of Noritoko Kyo Ki*, we read for July 14, 1408: "Karabe visited the temple and learned about the appearance of the guest star. Eight evil features were listed and reported to the authorities by the astronomical doctors. It is an astonishing happening." If the astonishing happening with its eight evil features was the same new star as seen in China, we now have an event of duration at least three months. Was it a supernova or merely a nova? If it was a conventional supernova, we would expect to see a remnant. There are two objects of interest in the region of sky where the new star occurred. One is an extended supernova remnant called CTB 80, and the other is a very famous binary X-ray source called Cygnus X-1. Both have been proposed by astronomers as appropriate remnants, although there are compelling reasons to reject both.

CTB 80 is a remarkable object, a plerionic remnant, with

giant protrusions. If the whole system was formed in a single outburst, it could not possibly have been as recently as 1408; thousands of years would have been needed for the protrusions to grow to their currently observed size. Perhaps the extended features are related to an ancient outburst, with the more recently formed plerion co-aligned by chance. This might be possible, but by now we are clutching at straws. It might be best to place CTB 80 aside as a fascinating and quite remarkable object but one which cannot be linked convincingly with the 1408 new star. We are forced to a similar conclusion with Cygnus X-1. This extensively studied binary system is believed to contain a normal massive star in circular orbit around a black hole. There is no evidence of a conventional extended supernova remnant surrounding Cygnus X-1, and the binary system as now observed is believed to be extremely old. Perhaps the 1408 new star was some sort of spurious outburst from Cygnus X-1, possibly initiated by a sudden increase in accretion rate. We can never hope to know with certainty. The 1408 new star may have been no more than a bright nova. If only the Ming dynasty astronomers had been as astute and diligent as their Sung forebears and left more detailed records, we might have been able to reach a more satisfactory conclusion.

The Sung astronomers hold the distinction of having recorded three supernovae—those of 1006, 1054, and 1181. It was to be almost four centuries before another supernova was sighted from Earth, by which time the standard of European observational astronomy had far surpassed that of the Far East.

8

RENAISSANCE SUPERSTARS

> Before the days of Kepler, the heavens
> declared the glory of the Lord.
>
> —*George Santayana*

Tycho Brahe was without doubt the greatest of the sixteenth-century (pretelescopic) observational astronomers. Tycho built astronomical measuring instruments of exceptional precision, and in 1576, with the financial backing of King Frederick II of Denmark, he constructed a palatial observatory—Castle of the Heavens—on the island of Hven in the Baltic to provide himself with a base for his observations. Much of Tycho's astronomical work was aimed at improving planetary observations so that he could obtain sufficiently good data to allow him to disprove or improve the Copernican model. In spite of a lifetime's meticulous astronomical records, he failed to achieve this, mainly because he tried to force them into a pre-Copernican, Earth-centered system. Despite this failure, Tycho's contributions to astronomy were truly epic, not least his work on the new star which shone forth in 1572 and which now bears his name.

Tycho was just twenty-six years old when the new star appeared. He would have been brought up and educated under a social system still based on Aristotelian beliefs and therefore opposed to the Copernican revolution. New stars were not yet part of the new cosmology of the era. Tycho was therefore unable to hide his amazement when he first saw the new star. At the time he was visiting an uncle at the monastery of Herritzuradt in Denmark. Tycho later wrote:

> When on the above mentioned day [November 11], a little before dinner . . . I was returning to that house, and during my walk contemplating the sky here and there since the clearer sky seemed to be just what could be wished for in order to continue observations after dinner, behold, directly overhead, a certain strange star was suddenly seen, flashing its light with a radiant gleam and it struck my eyes. Amazed, and as if astonished and stupefied, I stood still, gazing for a certain length of time with my eyes fixed intently upon it and noticing that same star placed close to the stars which antiquity attributed to Cassiopeia. When I had satisfied myself that no star of that kind had ever shone forth before, I was led into such perplexity by the unbelievability of the thing that I began to doubt the faith of my own eyes, and so, turning to the servants who were accompanying me, I asked them whether they too could see a certain extremely bright star when I pointed out the place directly overhead. They immediately replied with one voice that they saw it completely and that it was extremely bright. But despite their affirmation, still being doubtful on account of the novelty of the thing, I inquired of some country people who by chance were traveling past in carriages whether they could see a certain star in the height. Indeed, these people shouted out that they saw that huge star, which had never been noticed so high up. And at length, having confirmed that my vision was not deceiving me, but in fact that an unusual star existed there, beyond all type, and marveling that the sky had brought forth a certain new phenomenon to be compared with the other stars, immediately I got ready my instrument. I began to measure its situation and distance from the neighboring stars of Cassiopeia, and to note extremely diligently those things which were visible to the eye concerning its apparent size, form, color, and other aspects.

Most observational astronomers will have experienced Tycho's thrill and amazement at witnessing "that the sky had brought forth a certain new phenomenon." The heavens abound in apparently unique and enigmatic objects and strange phenomena. Rarely does a competent astronomer, with a carefully planned observational program, direct a big modern telescope at the sky without finding something exciting. Few, however, are given the opportunity to change dramatically the course of astronomical knowledge. Today, many such changes are the result of observers having the instrumentation capable of detecting, often unexpectedly, new phenomena—as happened with the discovery of pulsars. Other changes result from astronomers merely being in the right place at the right time. This was Tycho's good fortune, and it was astronomy's good fortune that an observational astronomer of Tycho's exceptional caliber was around to witness the 1572 event. Not that Tycho was the only good observational astronomer of the time—many others also took careful measurements of the new star. But Tycho had a certain flair and a fine expository style, plus a growing scientific reputation, which enabled him to present his observations in a way that made greatest possible impact on contemporary scholarship. Within a year, he had presented the results of his work on the new star in his preliminary report entitled *De nova stella,* a publication schedule that astronomers rarely better even today. His more detailed investigation of the nature of the new star was recorded in *Astronomiae instauratae progymnasmata,* published in 1602 shortly after his death.

Tycho did not discover the 1572 new star. He had had to cope with poor weather and noted, "Although several people had preceded me in seeing it, the air in our part of the world had not been clear enough." Discovery is often attributed to Wolfgang Schüler of Wittenberg, who recorded that he sighted a "comet" in Cassiopeia at about 6:00 A.M. on November 6. Schüler apparently did not have the courage to propose that the object was a new star, preferring to stay within the confines of Aristotelian dogma by proposing a cometary classification. It is possible, however, that Schüler was not in fact the first to sight and record the object. Francesco Maurolyco of Messina appended his account of the "comet" as follows: "Maurolycus, abbot of Messina, wrote this hastily, subject to the judgment of the more wise. November 6, 1572." Had Maurolyco seen the star before the date when he finally wrote down the sighting? Possibly, but not by more than a day or two. Certainly the star was not obvious on the night of November 2, and it was therefore presumably fainter than, say, about third magnitude, the brightness of many of the stars in Cassiopeia. The eminent mathematician Jerome Mugnoz of Valencia was definite about this point: "I am certain that on the second day of November 1572 there was not this comet in the sky, because on purpose more than an hour after 6 in the afternoon I showed in Hontinente to many people publicly how to identify the stars. . . ."

Even among those in Europe who did not support the cometary theory, the "newness" of the star was not necessarily accepted. Michael Maestlin of Tübingen, who himself made careful measurements of the star, noted: "Various judgments and opinions were heard about it from people who are not illiterate. Some conceded that it was natural and permanent, not new. . . . Others indeed believed that it was Vega, others Capella, others Arcturus; but others said Venus, others Saturn; others some other star or planet." Maestlin was later the tutor of Johannes Kepler, who had been borne the year before the new star of 1572 appeared. In later years, Kepler referred to the new star of 1572 as having heralded the advent of a great astronomer. This has usually been taken as his paying tribute to Tycho Brahe, unless he was rather immodestly making oblique reference to his own birth.

The fact that there could have been doubts as to the true identification of the new star this late in the Renaissance places in perspective the European failure to record the new star of 1054. The two new stars of 1054 and 1572 were of comparable brightness. If Renaissance scholars were still attempting to classify a "new" star as Venus or Vega, it is hardly surprising that their more conservative ancestors failed to record the birth of the Crab. Tycho seems to have adopted the attitude that the star was merely one of the permanent stars that God had chosen to unveil for the first time.

Tycho suggested that when he first saw the star, and throughout the rest of November, it emulated Venus when that planet is at maximum brightness, which would have been the case in November 1572. A direct comparison with Venus would have been possible in the early-morning hours. Other observers, however, doubted that it quite reached the brightness of Venus. Casper Peucer of Wittenberg noted that the new star was "brighter than all the planets and stars with the exception of Venus"; John Praetorius, also of Wittenberg, recorded that "it was larger and brighter than Jupiter, but easily fainter than Venus"; Maestlin estimated that it "surpassed Jupiter, and almost Venus." It seems that Tycho may have exaggerated slightly its brightness. At this time, Venus would have been of magnitude −4.3. A reasonable estimate of the new star's maximum magnitude would therefore seem to be −4.

Tycho followed in some detail the decline of the new star from maximum brightness. He described this as follows:

> When first seen, the nova outshone all fixed stars, Vega and Sirius included. It was even a little brighter than Jupiter, then rising at sunset, so that it equaled Venus when this planet shines in its maximum brightness. . . . The nova stayed at nearly this same brightness through almost the whole of November. On clear days it was seen by many observers in full daylight, even at noontime, a distinction otherwise reserved to Venus only. At night it often shone through clouds which blotted out all other stars. However, the nova did not retain this extraordinary brightness throughout its whole apparition but faded gradually until it finally disappeared. The successive steps were as follows: As already stated, the nova was as large as Venus in November [1572]. In December it was about equal to Jupiter. In January [1573] it was a little fainter than Jupiter and surpassed considerably the brighter stars of the first class. In February and March it was as bright as the last-named group of stars. In April and May it was equal to the stars of the second magnitude. After a further decrease in June it reached the third magnitude in July and August, when it was closely equal to the brighter stars of Cassiopeia, which are assigned to this magnitude. Continuing its decrease in September, it became equal to the stars of the fourth magnitude in October and November. During the month of No-

> vember, in particular, it was so similar in brightness to the nearby eleventh star of Cassiopeia that it was difficult to decide which of the two was the brighter. At the end of 1573 and in January 1574 the nova hardly exceeded the stars of the fifth magnitude. In February it reached the stars of the sixth and faintest class. Finally in March it became so faint that it could not be seen any more.

In this description Tycho was almost certainly making comparisons with the stars of given magnitude from the catalog of Ptolemy's *Almagest.* Apparently during the months of June and September 1573, for which no comparisons were made, there were no suitable reference stars.

From the above data, and with a few minor corrections and assumptions, it is possible to produce a light curve for the 1572 new star. This is shown as Figure 27. Included for comparison

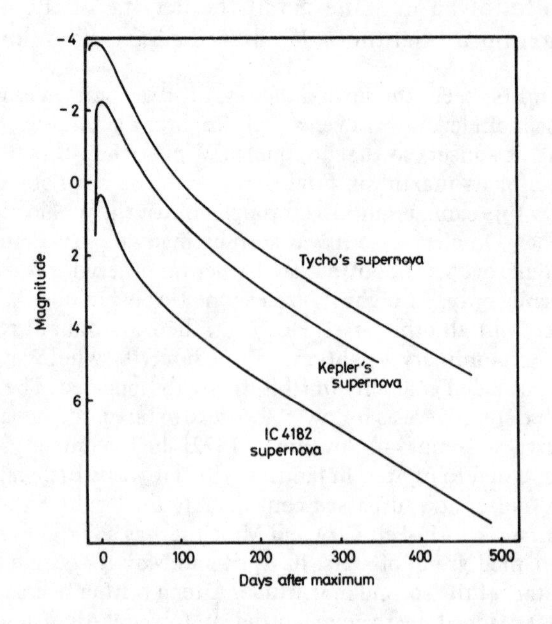

Figure 27. The light curves for the supernovae of Tycho and Kepler compared with the light curve of the Type I supernova in the galaxy IC 4182. (Magnitude scale refers only to the two Galactic supernovae.)

is the light curve of the supernova discovered by Zwicky in the galaxy IC 4182—the prototype Type I supernova. The similarity in form of the two light curves suggests that Tycho's supernova was undoubtedly of the Type I.

Far Eastern records of the supernova add nothing to the European records. Nevertheless, they may be quoted to show how static Oriental astronomy had remained since the Sung supernovae were recorded; during the intervening period, by contrast, European astronomy had been on the ascendant. In the astronomical treatise of the history of the Ming dynasty, there is but a terse record: "There are also some stars which did not exist in antiquity but which exist now. Beside T'sê-hsing [the 'Whip star'—see Figure 26] there is a guest star which newly appeared during [1572]. At first it was large, but now it is small." In the draft of the history, the star was incorrectly classified as a *hui-hsing* ("broom star"), though the error was corrected in the final history.

In the annals of Emperor Shên Tsung, a more detailed account is given:

> [November 8, 1572] . . . at night a guest star was seen in the northeast; it was like a crossbow pellet. It appeared beside Ko-tao [asterism—see Figure 26] in the degrees of Tung-pi [lunar mansion]. It gradually became fainter. It emitted light in the form of pointed rays. After [November 24] at night the said star was reddish-yellow in color. It was as large as a lamp and the pointed rays of light came out in all directions. . . . It was seen before sunset. . . . At the time the Emperor saw it in his palace. He was alarmed and afraid, and at night he prayed in the open air on the Vermilion Steps. . . . The star gradually diminished in brightness only in [March 1573]. When we come to [April/May 1574] it finally disappeared.

It seems that the Chinese astronomers followed the new star for a month longer than Tycho, which is a testimony to their diligence, possibly assisted by fairer weather. The allusion to Ko-tao, eight degrees from the new star's actual position (and not within Tung-pi lunar mansion), is rather surprising. Ko-tao was the missing asterism in the account of the 1181 star. Again the explanation for this may be astrological. Alternatively, since the annals of the emperor were not based on the official astronomical records, the nearer asterisms of the charioteer and his whip may not have been known to those who recorded the emperor's frightened response to the appearance of the new star. They did, however, give the correct lunar mansion. The true proximity of T'sê-hsing is confirmed in the single Korean record: "[November 1572] . . . A guest star was seen in the vicinity of T'sê-hsing. It was larger than Venus."

If, as for the earlier historical supernovae, we had been dependent on the Oriental records, we would have been able to classify the star as a supernova on the basis of duration (18 months) and extreme brightness, comparison with Venus plus daylight sighting making it of magnitude about −4. Its position would have remained uncertain, but the European observations, especially those of Tycho, Maestlin, and the English mathematician Thomas Digges of Cambridge, allow us to pinpoint the location with some precision. All three located the new star in relation to the polestar, the principal stars in Cassiopeia (Figure 28), and other nearby bright stars. Their chief

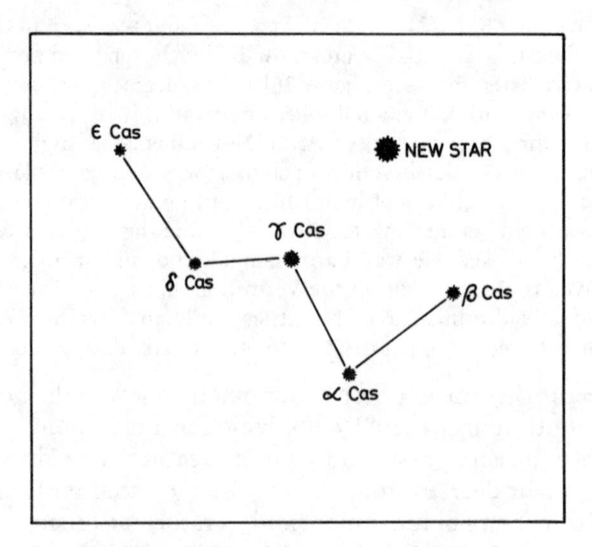

Figure 28. The position of the new star of 1572 relative to the principal stars in Cassiopeia.

aim in accurately measuring the position was to determine whether the "new star" belonged to the eighth sphere of fixed stars. Digges adopted a very simple procedure, which in principle was capable of yielding a very accurate result. He selected two pairs of reference stars so that, as accurately as could be judged, the new star lay at the intersection of the lines joining each pair. Digges checked the alignment with a long ruler and was able to conclude that the star kept a fixed position relative to the background stars, to within at least 2 arc minutes (there are 60 arc minutes to the more familiar angular measure of a degree). Maestlin adopted a similar technique to Digges, substituting a taut thread for Digges's ruler, and was able to reach a similar conclusion. Modern-day interpretation of the data of Maestlin and Digges allows the position of the new star to be found to within a few arc minutes.

The technique adopted by Tycho should have been very accurate. At the time the new star appeared, he had recently finished making a new sextant, graduated in minutes of arc and supposedly capable of measuring angles to an accuracy better than this. Obviously keen to make use of his new instrument, Tycho carefully measured the angular separation of the star from nine of the brightest stars in Cassiopeia and from the polestar. The data were included in his *Progymnasmata* in a table entitled "Distances of the new star from some of the principal fixed stars in the constellation of Cassiopeia, diligently investigated and verified." Tycho obviously had no doubts as to the precision and reliability of his measurements. Yet when these data are used to estimate a position for the new star, this is found to differ by about six arc minutes from the Digges/Maestlin position. Who was right? History tended to pass judgment in favor of Tycho Brahe. He was after all the more experienced observer. He was the builder of instruments of exceptional precision and in later life calculated the positions of over 700 bright "permanent" stars, positions that we now know were all accurate to better than two arc minutes. In general his technique was of much greater accuracy, and he later checked the angular separation of the nine principal stars in Cassiopeia (using a different instrument) with an accuracy of the order of one arc minute. In spite of this, we now know that it was Tycho's position for the new star that was in error, and the reason for this was his new sextant.

Evidence for such a judgment only came with the discovery of the remnant of the 1572 supernova, which was one of the nonthermal radio sources discovered in the early days of radio astronomy. High-resolution observations showed this object, usually referred to as 3C 10, to be a well-defined doughnut remnant. Later it was discovered to have faint optical filaments around the shell and later still to be a moderately strong X-ray source. The center of the remnant, which one would naturally suppose to be the site of the supernova outburst of 1572, was found to lie four arc minutes east of Tycho's position, which lies near the periphery of the remnant. Where had Tycho gone wrong? By fitting his data to the true position of the supernova (taken to be the center of the presently observed remnant), one finds that all his angular measurements show a systematic error of about two arc minutes. When allowance is made for this, his data give a position for the new star right in the center of the remnant. The cause of the systematic error is uncertain. Perhaps it was in part due to the way he aligned the instrument, and his inexperience of using it. Perhaps the new wood of the sextant had warped. We shall never know, since Tycho made no mention of ever having used the instrument again. Perhaps he suspected its imperfections, though he does not confess to this. He could never have guessed that four centuries later, giant radio and optical telescopes, and X-ray instruments on satellites, would reveal the positional error as they scanned the energetic remnant from his *nova stella.*

In 1596, after falling out with his patrons in the Danish Royal Court, Tycho Brahe left Hven and his native Denmark to become imperial mathematician to the Holy Roman Emperor Rudolf II. His new observatory was a castle near Prague. Tycho was soon joined by a young assistant, Johannes Kepler. After Tycho's death in 1601, Kepler took over his official position, and the task of interpreting the massive records was entrusted to him. Kepler's interpretation of the Hven records, and in particular the observations of Mars, gave a firm observational basis to the Copernican cosmology, modified by Kepler to accommodate elliptical orbits for the planets. Just as Tycho Brahe's astronomical career had been directed by the appearance of a new star, so it was to be with Kepler. Kepler's study of the supernova of 1604 ranks, with his study of the planetary motions, as one of the major contributions to the science of the age.

The 1604 supernova was discovered a full two weeks before its maximum, a quite remarkable time-span, not achieved for the brighter Galactic supernovae of 1054 and 1572. As noted earlier, it is possible that the spectacular 1006 supernova may have been discovered four weeks before its maximum. The early discovery was the result of a fortunate coincidence. A close conjunction of Mars and Jupiter, with Saturn relatively nearby also, had been predicted for early October in the southern sky. Many experienced observers were therefore eagerly searching the appropriate area of the sky. The new star appeared within the constellation Ophiuchus, and just three degrees to the northwest of the planets in conjunction. Anyone studying the conjunction could hardly have missed such a nearby event. Discovery in Europe was on October 9, when it was reported independently by an anonymous physician of Cosenza and Altobelli of Verona. David Fabricius and many other very experienced observers were insistent that the new star had not been present the previous night, when they had been observing the conjunction.

By October 10, Kepler had been notified of the new star's presence by a Prague courtier called Bronowski. Bad weather, however, had set in, so that he was not able to observe it until a full week later. He then subjected it to detailed investigation during the next twelve months, reporting his observations in *De stella nova in pede serpentarii,* a work of "astronomical, physical, metaphysical, meteorological, and astrological discussion." It is somewhat surprising to see a professional scientist of Kepler's standing reverting to astrological interpretations, though to what extent this was merely to please his patron is unknown. By producing a book on the 1604 supernova, Kepler ensured, as Tycho had done for the 1572 star, its permanent commemoration in his name. It must be remembered, however, that Kepler, like Tycho, was not the discoverer of his star, and many other outstanding observers left valuable information that has contributed to our present understanding of the event.

Using detailed comparisons with the brightness of the planets and certain stars, Kepler followed the rise and decline in brightness over many months. The probable date of maximum brightness was near November 1. As we shall see later, this estimate comes from Korean observations. There were no useful estimates of brightness from Europe between about October 17 and November 17. Kepler relocated the star after a period of it being a daylight object, on January 3, 1605. It was clearly too faint to be seen by day. On rediscovery, Kepler wrote that "it was much lessened from its original magnitude." He last saw it on October 8, 1605: "Now exactly a year after its first appearance in a very clear sky its appearance could only be noted with difficulty. . . ." Later attempts to sight it were unsuccessful. Although Kepler's reports of the brightness were far less frequent than those of Tycho for his star, it is nevertheless possible to reconstruct a light curve of the event. This is plotted in Figure 27. Like its predecessor, the new star was undoubtedly a supernova of Type I.

The standard of the three extant Chinese records of the star is low. They do reveal that it was first seen in China on October 10 and that it was "finally extinguished" on October 7 the following year, which gives a time span almost identical to the European observations. It was described as being "like a crossbow pellet" (obviously a favored description of the Ming historians—the same was used for Tycho's star) and reddish-yellow in color. There are no useful estimates of brightness, nor is an accurate position given. Compared with the quality of earlier Chinese new-star sightings, the records are disappointing. It is impossible to judge whether this reflected a lack of astronomical interest or an overstrict editing by the chroniclers. By contrast, the regularity and detail of the Korean observations surpass those for any other Oriental new-star sighting. On almost every clear night, the Korean astronomers reported the star's position, its color, and its brightness compared with neighboring stars and planets. We are regularly told whether thick cloud ever blanketed the sky, ruling out observations. In all, almost one hundred observations were recorded, forty of these being affected by cloud or the Moon. This day-to-day diary of observations is unequaled from anywhere else in the world up to this time.

The first report from Korea reads: "[October 13, 1604]—In the first watch of the night a guest star was seen in the tenth degree of Wei [lunar mansion] and distant 110 degrees from the [north] pole. Its form was slightly smaller than Jupiter. Its color was yellowish-red, and it was scintillating." Apart from changes in brightness and a slightly revised position, subsequent records are similar. Between October 28 and November 5 the star was compared with Venus; furthermore, from October 28 to 31 "its ray emanations were very resplendent," which suggests that this was the period of maximum brightness. From December 4 to 27 the star was a daylight object from Korea: "The guest star was close to the Sun. It sank in the west before dusk, and could not be observed." Regular reports and comparisons followed its reappearance, until a final reference dated September 15—three weeks before the keen-eyed Chinese and Europeans lost sight of it.

The Korean data alone would allow an independent light curve to be constructed. The comparisons with Venus near maximum were probably unreliable, since the two could not be seen concurrently at this time. On the other hand, it must have been significantly brighter than Jupiter, making it close to magnitude −3 at maximum (with an uncertainty of at least half a magnitude). The similarity of the Korean and European light curves is a remarkable testimony to the accuracy of pre-telescope photometric observations from two widely separated scientific cultures.

One Korean observation allows an accurate position for the new star to be estimated. On January 20, 1605, it was recorded that "Venus invaded [*fan*] the guest star." We learned earlier that *fan* meant an approach of less than about a degree. The precision of the term is again emphasized by modern calculations of the separation between the known position of the new star and the planet on the day in question—a mere 0.51 degrees. In the absence of the European measurements, this single observation would probably have been sufficiently accurate to allow the remnant of the new star to be identified with certainty. In any case, thanks to the efforts of Kepler and Fabricius, this is merely an academic exercise. The Europeans measured the angular separation of the supernova from several reference stars in the vicinity. These measurements were of exceptional precision, showing remarkable internal consistency. They allow us to position the new star to better than a minute of arc and allowed Kepler and Fabricius to locate the nova among the "fixed stars" and deliver a death blow to Aristotelian cosmology.

An accurate position for the new star allows its remnant to be identified with certainty. Centered on Kepler's *stella nova* position is a nonthermal radio shell called 3C 358, within which lie isolated, bright, optical knots. The X-ray remnant was discovered with HEAO-A and has now been accurately mapped from the *Einstein* Observatory.

The failure to detect the supernova that produced the spectacular remnant known as Cassiopeia A remains something of a mystery. The radio remnant, at a distance of about 8,000 light-years, was one of the earliest discoveries of the nascent science of radio astronomy in 1948. Optical nebulosity was located at the site in 1954. The optical remnant has two distinct components; the so-called quasi-stationary foculi (QSF), showing no evidence of rapid expansion and believed to represent material lost by the star before it exploded, and the high-velocity knots (HVK) depicting stellar debris showering outward at thousands of kilometers per second from the actual supernova outburst. By extrapolating backward the expansion of the high-velocity knots, a date for the supernova of about 1670 has been determined. By such a time we were well into the telescopic era of new astronomical enlightenment. Astronomers of the caliber of Hevelius, Flamsteed, Huygens, Hooke, Boyle, Cassini, Halley, Roemer, plus many others, were assiduously and methodically studying the heavens.

Cassiopeia is one of the most famous and best-known constellations; it is circumpolar from Europe, so is observable throughout the year. If the seventeenth-century supernova in Cassiopeia was normal, at least insofar as we now believe we understand what a normal supernova is, could it have gone unnoticed? Although this region is now known to be heavily obscured, calculations show that the outburst should have approached the zero magnitude, making it the brightest object among the bright second- to fourth-magnitude stars in Cassiopeia. It seems inconceivable that it could have been as brilliant as this and escaped notice by the talented "new-wave" stargazers. Indeed, there seems to be no escaping the conclusion that it must have been a very much fainter event than a normal supernova, and theories have been developed involving a massive star losing much of its outer envelope (hence the QSF). At outburst the light-emitting photosphere would be smaller than usual for a supernova, and the silent supernova resulting is intrinsically very much fainter than normal. But even allowing for this possibility, the seventeenth-century progenitor of Cassiopeia A must have brightened to about sixth magnitude, and one candidate for a detection at this level does exist.

The Royal Observatory at Greenwich was established by royal decree of King Charles II in 1675. Its early aim was not specifically to advance the science of astronomy but to foster observations leading to an improvement of navigation. The problem of accurately determining longitude at sea still awaited solution. The king appointed the self-taught but highly accomplished young astronomer John Flamsteed "our astronomical observator, forthwith to apply himself with the most exact care and diligence to the rectifying the table of the motions of the heavens, and the places of the fixed stars, so as to find out the so-much-desired longitude of places for the perfecting of the art of navigation." "Exact care and diligence" became Flamsteed's hallmark, to the point where he was severely criticized by the scientific establishment of the time (under the dominating influence of Newton) for not publishing his results. Rather, he wished to subject his data to the most detailed scrutiny and checking with the intention of eventually publishing a new British catalog of stars far surpassing in reliability and accuracy the still-accepted catalog of Tycho. The story of how Flamsteed's dream was frustrated by the intervention of Newton is well known and is not atypical of the enmity generated within the astronomical community even today, when the goals of individual researchers and scientific administrators differ. Flamsteed was forced to hand over his incomplete observations to the Royal Society. Despite assurances that the catalog would not be published without Flamsteed's consent, Edmund Halley (later his successor at Greenwich) edited a version published in 1712. By 1715 Flamsteed had recovered most of the copies and had them destroyed, "that none might remain to show the ingratitude of two of his countrymen [Newton and Halley]." John Flamsteed, always of poor health, died in 1719 before finishing his masterpiece; it was completed by two of his assistants, Crosthwait and Sharp, as *Historia Coelestis.* Its publication in 1725 established Flamsteed as the greatest observational astronomer of the age.

One fascinating entry in the first volume of *Historia Coelestis* presents sextant observations of a star designated "Supra τ." On the night of August 16, 1680, Flamsteed measured the angular distances of this star he estimated to be of sixth magnitude from two reference stars called Scheat (β Pegasi) and Algol (β Persei). The reduced observations of Supra τ are entered in the 1725 catalog as the star 3 Cassiopeia. But at the position given for 3 Cassiopeia, no star now exists! The inconsistency was recognized by Caroline Herschel toward the end of the eighteenth century, but she concluded that Flamsteed must have made an uncharacteristic error. Halley must have reached a similar conclusion ninety years earlier, since 3 Cassiopeia was not included in the ill-fated 1712 catalog. However Flamsteed and his co-workers clearly had faith in the original observations, and 3 Cassiopeia was reinstated in the 1725 catalog.

So what did Flamsteed observe on that summer night in 1680? Perhaps he simply made an error, since it has been pointed out that the two angular distances he recorded could correspond to two different stars, rather than a single object at the intersect of the arcs. The arc from Scheat agrees with the position for a star now called AR Cas within an error of one-half minute of arc, and the arc from Algol coincides with a star called BD +56° 2999 within an error of about one minute of arc. But such an error was not like Flamsteed. Out of almost three thousand stars in *Historia Coelestis,* there are only four anomalous observations, including 3 Cassiopeia. But perhaps 3 Cassiopeia is not anomalous after all, since its position lies within about 10 minutes of arc from the center of Cassiopeia A. Such an error would be uncharacteristically large for Flamsteed if 3 Cassiopeia really was the supernova, although it should be remembered that his early sextant observations were not as accurate as his later (post-1689) mural arc observations. In addition, if Supra τ was a transient event, he would not have been able to make later observations to check his accuracy. Flamsteed could not have suspected that he was witnessing a remarkable cosmic phenomenon; he was later to observe Uranus in 1690, 1712, and 1715 without realizing it was a new planet, leaving the accredited "discovery" to William Herschel.

It would be nice to think that perhaps the faint birth pangs of Cassiopeia A did not go unheeded after all, but were recorded by the Reverend John Flamsteed as his object Supra τ. Sadly we will never know for certain.

Setting aside the special case of Cassiopeia A, no normal supernova has been observed in our galaxy since Kepler's new star. Hence the importance of the pretelescopic historical records of the Galactic superstars.

9

THE SUPERSTAR SOLUTION

He who asks questions, cannot avoid the answers.
—*Cameroonian Proverb*

THE SIX GALACTIC supernovae from the past two millennia for which probable or definite remnant associations have been made are summarized in Table 2. In addition, there is the certain supernova of 393, for which a definite remnant identification is unfortunately not possible, and the silent supernova of about 1680 proposed for the birth of Cassiopeia A.

Seven certain normal supernovae over a 2,000-year period might seem to imply one Galactic outburst about every 300 years. However, in the solar neighborhood, radio observations, which are not affected by the layer of interstellar dust concentrated about the Galactic plane that hides much of the Galaxy from view, indeed, show that supernova remnants are fairly evenly distributed over the face of the Galaxy, like cherries in a pie. Figure 29 indicates that the historical supernovae were confined to about a one sixth portion of the pie, in the solar neighborhood, while supernovae in the remaining five sixths could not be seen from Earth. If, therefore, in our one-sixth segment of Galactic-pie supernovae have occurred at least once

TABLE 2

The Historical Supernovae

Year of Discovery	*Apparent Magnitude at Maximum*	*Duration*	*Remnant Name*
185	−8	20 months	RCW 86 (probable)
1006	−10	2+ years	PKS 1459–41 (definite)
1054	−5	22 months	Crab (definite)
1181	0	6 months	3C 58 (probable)
1572	−4	18 months	Tycho's (definite)
1604	−3	12 months	Kepler's (definite)

Remnant Name	*Remnant Properties*			
	Radio	*X-rays*	*Optical*	*Neutron Star*
RCW 86	Shell	Yes	Yes	No
PKS 1459–41	Shell	Yes	Yes	No
Crab	Plerion	Yes	Yes	Yes (pulsar)
3C 58	Plerion	Yes	Yes	Yes (X-ray source)
Tycho's	Shell	Yes	Yes	No
Kepler's	Shell	Yes	Yes	No

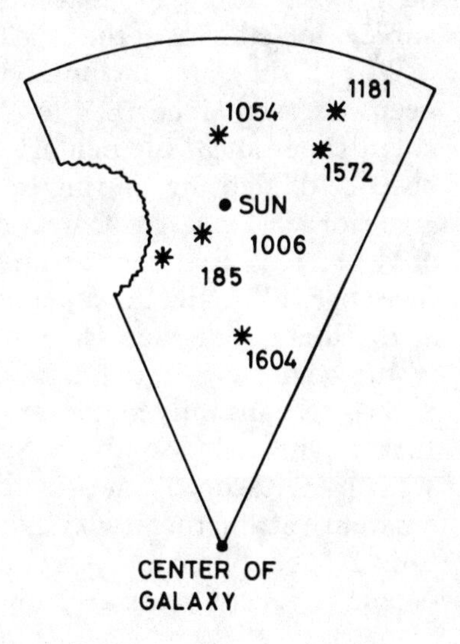

Figure 29. The position of the historical supernovae, with respect to the Sun and the center of the Galaxy. The "bite" shows that region too far south to have been observed by northern-hemisphere civilizations.

every 300 years on average, then within the whole Galaxy they must occur at least once every 50 years on average. This calculation, however, is based on the unlikely assumption that all the supernovae occurring within the one-sixth segment over the past 2,000 years were observed.

Some supernovae would have been daylight objects, not bright enough to be seen by day, and probably undetectable by the time they were again nighttime objects. Others would lie in the southern portion of the Galaxy, below the horizon from the careful astronomers of China. Almost certainly some historical new-star records, particularly from the first millennia of the new era, have been lost. Further, the Oriental histories contain only what the official historians saw fit to include, although it is unlikely that any spectacular new stars were overlooked. In any event, the assessment of historical new-star records is necessarily rather subjective. We can do little better than suggest that as many as half the supernovae within our segment during the past two millennia may have been overlooked for one reason or another, giving a final estimate for the characteristic, or average, interval between Galactic supernovae of, say, 20 to 30 years, with a large element of uncertainty. Such a value is close to that implied by extragalactic supernova surveys for galaxies of the type the Milky Way is believed to be.

Why, if they are so common, have no Galactic supernovae been observed since 1604 (or 1680 if one credits Flamsteed's silent supernova)? Elementary astronomy books often use the absence of sightings during the past several hundred years to argue for a much larger characteristic interval, such as hundreds of years. This argument assumes that, with the advent of the telescope, all Galactic supernovae would reveal themselves. Sadly, this is not the case. Our serving of Galactic pie today would be not very much larger than for pretelescopic astronomers—perhaps only a quarter portion. Indeed, the only factor that might increase the possibility of detection compared with pretelescopic times is the large number of competent amateur astronomers who assiduously monitor the skies for unusual phenomena. Several novae have been detected by such people in recent years at below the naked-eye limit. Big telescopes are never used for surveying the Galaxy for new stars. All in all, therefore, although we certainly seem to be overdue for the next Galactic supernova, its nonappearance does not seriously call into question the estimate of the characteristic interval from the historical supernova records. There are gaps of many centuries between the sporadic appearance of the historical supernovae, and the passage of almost four centuries since the last detected Galactic supernova seems not untypical. Most professional astronomers would hope to be near a telescope with suitable instrumentation when the next Galactic supernova does blaze forth. The realization that the bursts of light from many hundreds of supernovae that have already taken place within the Galaxy are presently on their way to us gives them some hope that it may be during their lifetime.

If supernovae occur about every 20 to 30 years on average throughout the whole of our galaxy, how often do they occur in the solar neighborhood? We must first decide on the possible sphere of influence of the expanding remnant of a supernova. Since a remnant reaches its extinction phase (after which it merges with the interstellar medium) in about 100,000 years and when its radius has reached about 100 light-years, only supernovae occurring closer than this distance from the solar system could be of concern. We therefore need to consider the frequency of just those supernovae occurring within a sphere of radius 100 light-years centered on the Sun; that is, to a volume of four million cubic light-years. The volume of the Galaxy, or at least that portion to which supernovae are believed to be confined, is about eight trillion cubic light-years.

If one supernova occurs, say, every 25 years in a volume of eight trillion cubic light-years, then in a volume of a mere four million cubic light-years one supernova might be expected to occur on average about once every 50 million years. Obviously, in such a calculation a number of simplifying assumptions have had to be made, but the above estimate gives a value likely to be correct at least in order of magnitude.

A time interval of some 50 million years between supernovae occurring close enough for their expanding remnants eventually to engulf the Earth may seem so long as to be of minimal concern. But measured against the extreme age of the solar system, such events are relatively frequent, having occurred probably about 100 times since its birth. We therefore are forced to conclude that supernovae probably do occur close enough to the Earth sufficiently often to have had a possible influence on its evolutionary history.

No nearby stars are likely to end their lives as supernovae within the next few million years. But what could the global impact of a nearby supernova be? Certainly such an outburst would be an extremely brilliant spectacle. But could its influence extend beyond the astonishment, bewildered curiosity, and admiration of those who observed it?

One possible mechanism by which a supernova could affect the Earth accompanies the outburst itself. Theoreticians estimate that, at the time of the explosion, up to one hundredth of one percent of the energy is contained in a short-duration burst of X rays and gamma rays. This estimated fraction of the total energy is so small that it has not yet been possible to verify theory by the experimental detection of such bursts from extragalactic supernovae, although future X-ray satellites may be sensitive enough to detect them. In the absence of positive experimental proof for the existence of the X-ray/gamma-ray bursts, it might seem a wasted exercise to consider their possible consequences. Nevertheless, the same theoretical models that predict the bursts make other predictions that have been successfully tested by observation, forcing astronomers to concede that a short-lived X-ray/gamma-ray burst probably is part of the supernova phenomenon.

The X-ray/gamma-ray burst from a nearby supernova would be almost completely absorbed in the Earth's upper atmosphere, so it would have minimal direct impact at the surface. On the other hand, the influence on the chemistry of the upper atmosphere is expected to be significant. The burst would encourage the nitrogen, which is a minor constituent of the upper atmosphere, to combine with oxygen to form nitric oxides. Nitric oxides normally exist in the upper atmosphere only in very small amounts. An increase in the concentration of the atmospheric nitric oxides would lead to the destruction of the

Earth's protective ozone layer. Ozone (O_3) is molecular oxygen in which each molecule contains three oxygen atoms rather than the two of normal molecular oxygen (O_2). The nitric oxides catalytically break down the O_3 ozone molecules to O_2. By this mechanism the nitric oxides usually present in the upper atmosphere help to maintain the ozone concentration at a particular level. If there were to be a large increase in nitric oxide concentration following the arrival of the X-ray/gamma-ray burst from a nearby supernova, it has been estimated that the concentration of the ozone layer would be depleted to just a few percent of its normal value.

Ozone is familiar to most people as a pale-blue gas with a peculiarly pungent smell, which may be formed by the action of electrical discharge—the first indication that a piece of electrical equipment may be faulty is often given by the smell of ozone. In the belief that ozone represented some form of enriched oxygen, ozone generators used to be used as air purifiers, supposedly having an invigorating effect on users. Such mistaken ideas have been corrected in recent years by a growing awareness of the true importance of atmospheric ozone. Various types of aerosol pressurizing agents that could contribute to the reduction of atmospheric ozone concentration (most notably the fluorocarbons) have recently come under close scrutiny, as have the pollutants from high-flying supersonic jet transport aircraft such as the Concorde. What is the reason for the sudden interest in protecting nature's delicate balancing act in maintaining the concentration of the ozone layer?

The powerhouse of life on Earth is of course the Sun. The Sun emits energy over a broad spectrum, from long-wavelength radio waves and infrared radiation through visible light to the shorter-wavelength ultraviolet radiation and X-rays. As previously explained, only visible light, and certain wavelength radio waves and infrared radiation penetrate the atmosphere to reach ground level. Ultraviolet radiation is almost totally filtered out by the ozone layer. For at least as long as the Earth has had such a protective ozone layer, life has evolved protected from the solar ultraviolet radiation, so that any increase in its intensity represents a departure from normality that may have profound significance for living systems. A relatively modest increase in ultraviolet flux is expected to result in an increase in the incidence of skin cancer. Furthermore, for certain species with calcified skeletons, an increase in the production of vitamin D induced by ultraviolet radiation may be toxic. A major increase in ultraviolet flux following the almost total destruction of the ozone layer would undoubtedly have even more calamitous consequences.

The genetic information of life is stored in the complex helical chain molecule DNA (deoxyribonucleic acid). The study of the structure of DNA and the deciphering of the genetic code in this hereditary material have represented one of the major breakthroughs of twentieth-century science. It is now known that ultraviolet radiation can destroy the essential structure of DNA. Nature has allowed for an acceptable spontaneous mutation rate in hereditary information, and indeed such mutations have contributed to possible adaptations to environmental changes. Nevertheless, a sudden change in the level of ultraviolet radiation is likely to lead to a mutation rate that no living species could accommodate.

It must be admitted that the ozone layer would recover very rapidly after any destruction induced by the X-ray/gamma-ray burst from a nearby supernova. The burst itself is short-lived (of the order of 1,000 seconds) and is a one-time event for each supernova. The recovery time of the ozone layer is just a few tens of years. Life on Earth could be expected to recover quickly from the harmful mutations imposed on a single generation. A nearby supernova, however, will not yet have completed its dastardly task. The X-ray/gamma-ray burst travels to Earth at the speed of light, so that in the case of a supernova at a distance of 10 light-years, for example, the burst reaches Earth 10 years after the explosion. By contrast, the expanding shell of ejecta, traveling initially at a few thousand kilometers per second, will take over 1,000 years to reach the Earth. The Earth will then remain embedded in the expanding shell for many hundreds or even thousands of years. High-energy cosmic rays trapped in the expanding shell will almost certainly influence the ozone layer indirectly in exactly the same way as the initial X-ray/gamma-ray burst. But now the destruction of the ozone layer and the resultant catastrophic effects on Earth's life-forms will have a long-term effect. Prolonged exposure to increased ultraviolet radiation could well be disastrous for certain species. This has been suggested as the possible cause of the extinction of the dinosaurs at the end of the Cretaceous period, some 63 million years ago, although of course there are many other (perhaps more plausible) explanations for their demise.

It is not only the biological history of the Earth that may have been affected by nearby supernovae, for these may also account for catastrophes in the climate. Several forms of extraterrestrial influence have been suggested as possibly having influenced the Earth's climate. Long-term weather patterns on Earth are recognized as being highly variable. Various indicators of past climate, such as information contained in tree rings, ice cores, deep-sea sediments, and other geological samples suggest that over the past several hundred million years, climate has shown major excursions. These cover a wide variety of time scales, from the great ice ages of the geological past to short-term excursions of historical times. Examples of the latter are periods of extreme cold such as the "Little Ice Age" of the seventeenth century or the prolonged drought conditions experienced in medieval Europe. Furthermore, during the past several hundred million years there have been perhaps as many as six mass extinctions of major biological species, the best known being the dinosaurs. We cannot at present escape the conclusion that these extinctions may well have been a consequence of climatic excursions.

The global climate system displays great complexity, and despite significant advances in recent years, the present scientific understanding of the system is still only rudimentary. The reasons for climatic variability and change are not well known but are undoubtedly manifold. It is clear that potential sources of variability exist within the climate system itself, the principal components of which are the atmosphere, the oceans, the land, the biosphere, and the cryosphere (the Earth's ice and snow cover). Furthermore, several processes external to the climate system have been proposed as the causes that might force change. These processes are environmental events whose occurrences are independent of the state of the climate system itself. For example, possible influences of a terrestrial origin are volcanism, various human influences on environment (for example, deforestation, burning of fossil fuels, use of agricultural fertilizers, and so on), and tectonic processes in which the continents are alternately dispersed and reclustered. Volcanic eruptions may inject veils of fine ash into the stratosphere, covering a large geographical area and lingering for many years. The stratospheric ash veil intercepts incident radiation from the Sun, thus bringing about thermal gradients in the atmosphere, causing new atmospheric circulation patterns, and consequently affecting climate. Over the time scale of hundreds of millions of years, tectonic processes may affect climate through their influence on the depth and shape of the oceans, the extent of the ice cover, the height and positions of the continents, as well as other geophysical conditions to which the state of the climate system is sensitive.

So what might be the climatic impact of a nearby supernova? The production of nitric oxides in the upper atmosphere described earlier would create a global nitric oxide smog. The resulting decrease in the solar radiation reaching the surface of the Earth would lead to a fall in photosynthesis and the collapse of food chains. Furthermore, the ozone layer might be effectively removed, possibly for centuries, while the Earth remained embedded in the shell of supernova material. Since absorption of solar ultraviolet radiation by the ozone layer is an important mechanism for heating the stratosphere, the destruction of the ozone layer brought about by a supernova would remove the heat source. On the basis of what we know about heat transfer, there would be a subsequent small drop in temperature at the Earth's surface, not by itself sufficient to initiate a major climatic excursion but likely to be accentuated by dynamical coupling of the stratosphere with the underlying troposphere and lower atmosphere. The climatic excursion resulting might not be a full-blown ice age, but it could nevertheless be significant.

Considerations of the climatic and biological impact of nearby supernovae are very speculative, and there is no firm, direct evidence for any of the consequences suggested here. Speculation is, however, a legitimate part of science, so long as exaggerated claims are not made on the basis of conjecture. As Mark Twain noted: "There is something fascinating about science. One gets such wholesale returns of conjecture out of such a trifling investment of fact." The global impact of supernovae remains mere conjecture.

Earlier, we discussed when and in what time intervals supernovae occurred. We can now try to discover where and why they took place and what type of supernovae they were.

Extragalactic surveys tell us several things about the behavior of supernovae in spiral galaxies. The supernovae tend to occur close to the plane of the parent galaxy. Types I and II are equally likely to be seen, with Type II in particular usually occurring within the spiral arms of the galaxy. The arms of a spiral galaxy are believed to reveal a self-sustaining "density wave" through which the material of the galaxy (stars and gas) passes as it rotates like a giant Catherine wheel. Some stars, astronomers believe, are born as clouds of gas collapse on crossing compression lanes on the leading edge of the spiral arms. Massive luminous stars are expected to evolve comparatively rapidly, within less than about 10 million years, and explode as Type II supernovae before leaving the spiral arms in which they are formed.

The less massive stars take much longer to evolve than their larger, short-lived relatives that are born and die within the confines of the spiral arms. The comparatively small stars are expected to orbit a spiral galaxy many times during a lifetime and deviate from the plane of the galaxy before sedately evolving (after undergoing some protracted mass loss) into a white dwarf. A white dwarf in a binary system is then, of course, a potential site of a supernova of Type I if successive accretion drives it beyond the Chandrasekhar limit. According to the above argument, Type I supernovae would not be expected to be confined to the arms of spiral galaxies but should occur randomly throughout a galaxy. The association of Type I supernovae with old star populations (the so-called Population II), and Type II supernovae with young star populations (the so-called Population I) has been realized since the early days of extragalactic supernova surveys. These associations have been reinforced as the number of supernovae discovered has increased. Since Type II supernovae come from massive stars, they might be expected to be absent from elliptical galaxies, which are believed to be populated by old, low-mass stars. In fact, all the supernovae so far discovered in elliptical galaxies have been of the Type I. However, the theory that Type II supernovae are produced solely by comparatively young massive stars, while all Type I supernovae come from the comparatively old, less massive stars, has been called into question. Statistical studies suggest that Type I supernovae occur more frequently in galaxies displaying young stellar populations and signs of intense star formation, implying that a significant proportion of the Type I's might come from among the short-lived stars. The conflicting evidence on the progenitors of Type I supernovae remains to be resolved, although there seems to be a growing belief that a Type I outburst involves total disruption of a star without the formation of a neutron star.

From the birthrates for supernovae derived earlier, it is possible to decide the mass range of stars exploding as supernovae. This argument is presented as Appendix 16, where it is concluded that *all* stars between about 8 to 20 solar masses explode, as Type II supernovae, leaving a neutron star; but only about one in a thousand stars smaller than 8 solar masses explodes, as Type I supernovae, without leaving a stellar remnant.

One remaining problem in estimating birthrates relates to pulsars. Some statistical studies seem to imply a rather higher birthrate for pulsars than derived above for supernovae. Estimates as high as one pulsar born every five years have been suggested. Yet a one-to-one correspondence has been proposed between Type II supernova outbursts and neutron-star formation. This assumption may yet prove to be incorrect. It is to be hoped that future investigations will bring the discrepant rate for pulsars into closer agreement with the supernova rate of about one every 50 years for Type II's; it may still be possible that the implied high birthrate for pulsars is real, and that they have an alternative, and as yet unrecognized, formation mechanism not involving optical supernovae.

Because a pulsar's emitted radiation is concentrated into a narrow beam, the Earth would be expected to lie within the path of the sweeping beam of only about one in every four pulsars (although some neutron stars might still be observed in X rays—for example, the X-ray point source in the center of RCW 103 is believed to depict emission directly from the cooling neutron star, which is *not* seen as a pulsar). It is therefore not surprising that only one of the six remnants of historical supernovae listed in Table 2 has a detected central pulsar, although one other has a point X-ray source. Nor is it surprising that the vast majority of pulsars do not lie at the center of supernova remnants. Most of the cataloged pulsars are believed to be comparatively old objects, with ages of the order of many hundreds of thousands or even millions of years. By contrast, within a few tens of thousands of years most of the extended remnants that have survived to middle age are expected to merge with the interstellar medium and be unrecognizable. Hence, pulsars older than, say, 100,000 years will certainly have lost the extended remnants that presumably once surrounded them.

One further observational result remains to be considered: few of the cataloged 130 Galactic radio remnants have pulsars at their centers. This is partly the effect of beaming, but it is also due to the problem of detecting distant pulsars. The individual pulses from a pulsar tend to be spread out—they are said to undergo dispersion—during their passage through the interstellar medium. This effect can be partly compensated at detection, but if a pulsar is very distant, individual pulses will have become so spread out by the time they reach the Earth that they eventually merge, with the result that the pulsar can no longer easily be identified by its pulsed emission. Despite this observational problem, most pulsars within about 20,000 light-years should still be detectable. At least 30 extended supernova remnants lie within such a range. If we assume that half these remnants were formed by Type II supernovae and allow a factor of four for beaming, it seems that about four of these nearby supernova remnants should have pulsars at their centers. In fact, three such associations have been noted—the Crab Nebula, the Vela remnant, and the object known as MSH 15-52.

Pulsars are known to be high-velocity objects traveling at speeds of several hundreds of kilometers per second. There are two possible mechanisms for achieving such incredible speeds, which are equivalent to over a million kilometers per hour. The first is that the supernova explosion that created the pulsar may have been nonsymmetric—in other words, ejected material may have been thrown off in the explosion preferentially in one direction, the pulsar flying off in the opposite direction. The second explanation is that the presupernova system was a binary with two stars orbiting each other. The supernova explosion then disrupted the binary, with both the stellar object and companion star being thrown clear, in opposite directions, at high speed. If this second explanation is correct, then for every high-velocity pulsar, one would expect to see a high-velocity ex-companion traveling in the opposite direction. The Galaxy is full of high-velocity stars, the origin of which was considered unknown until recent years, although several decades earlier Zwicky and Baade had proposed a supernova origin. Many of the high-velocity runaway stars are now understood to be the "jilted" partners of previously close stellar relationships (see Figure 30).

Since pulsars are runaway objects, some would escape the confines of the gradually decelerating expanding supernova remnant within about 50,000 years. Several pulsars have been found close to the periphery of old extended remnants, and a few uncertain associations have been proposed.

Not all presupernova systems that are binaries are disrupted in the explosion. For some, the neutron star created in the supernova remains bound to its companion; these systems may eventually evolve to the compact X-ray objects already described. About 10 percent of binary systems that undergo a supernova explosion are expected to remain bound. Since

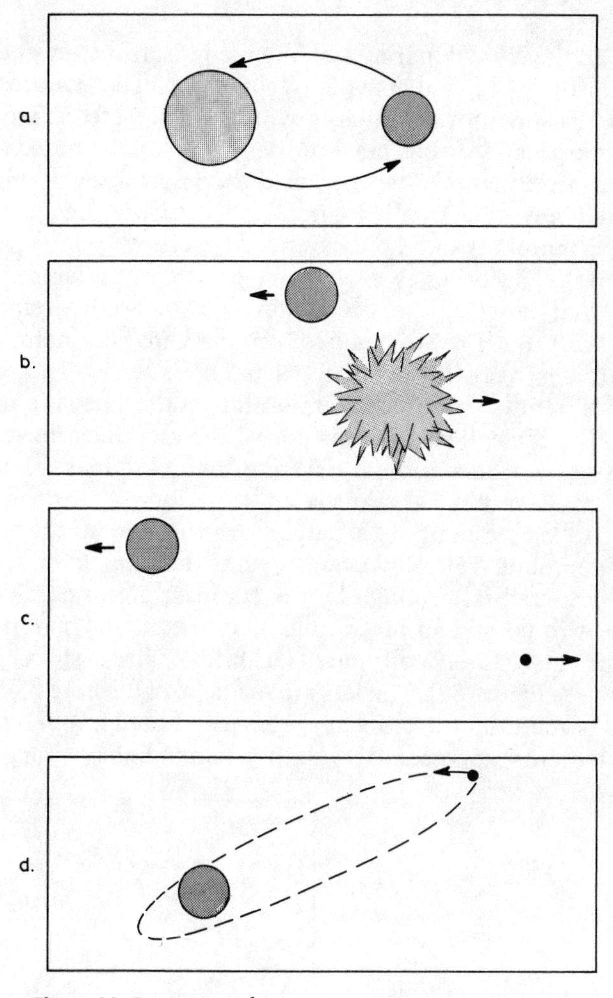

Figure 30. Runaway pulsars.

a. A binary system.
b. A supernova explosion of the more massive partner.
c. The neutron-star remnant of the supernova travels in one direction, and the less massive binary companion in the opposite direction.
d. In certain circumstances, the neutron star remains in a highly elliptical orbit around its companion, and the complete binary system travels away from the site of the supernova explosion.

nearly all the X-ray binaries are, like the pulsars, believed to be comparatively old, and also high-velocity objects, few associations with supernova remnants would be expected. Two unusual compact X-ray objects, however, may be associated with supernova remnants. The first of these is an intense, highly variable X-ray object called Circinus X-1, which lies just beyond the periphery of an old extended supernova remnant (see Figure 31). The other is a very much weaker variable X-ray source that lies close to the center of a supernova remnant called W50. Both these compact objects show dramatic outbursts (flares) at radio wavelengths and have both been associated with similar peculiar emission-line stars. The star at the center of W50, called SS433, is one of the most bizarre stellar objects known. It is a binary system involving mass transfer from a massive normal star via an accretion disk to the collapsed stellar remnant of the supernova that produced the extended remnant W50. However, it seems that the material cannot be accreted fast enough by the compact object (a neutron star or even possibly a black hole), and the excess material is shot out in two finely collimated jets from opposite sides of the system (see Figure 32). The jets are made up of discrete blobs or bullets of material traveling at speeds (estimated from Doppler measurements) approaching a quarter that of light—thousands

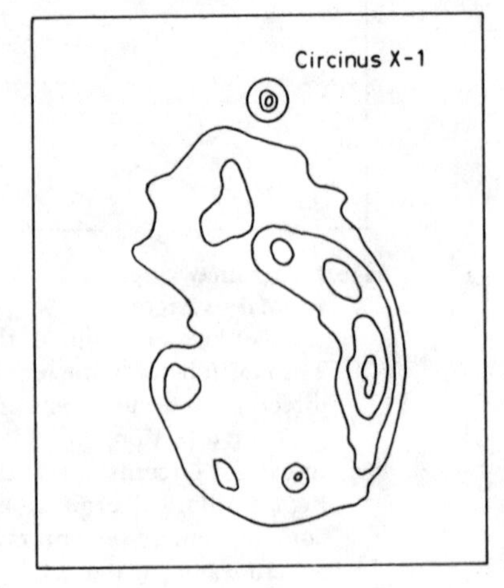

Figure 31. A radio map of Circinus X-1 and the supernova remnant G321.9-0.3.

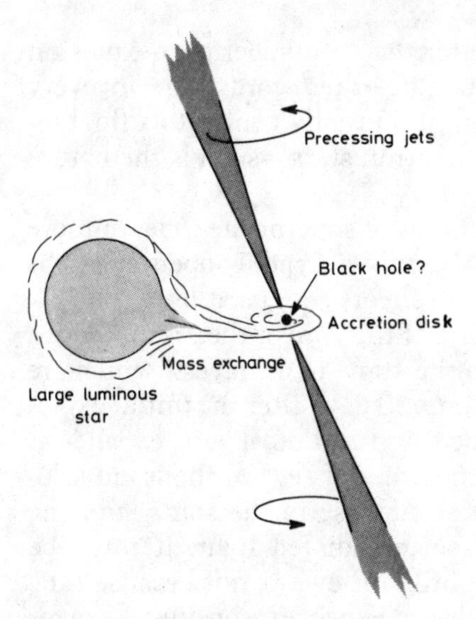

Figure 32. The bizarre object SS433.

of times faster than a bullet from a gun and faster than any material previously detected in the Galaxy. The jets extend for distances of many light-days and precess, like a giant garden sprinkler, with a period of 164 days. Just how the excess accreting material is collimated into fine jets and accelerated to extreme speeds remains uncertain. What is certain is that this enigmatic remnant of an ancient superstar will remain the object of intense observation and theoretical speculation.

It is a tantalizing possibility that Circinus X-1 and SS433 may represent a reasonably common manifestation of the stellar remnants of supernovae. If they are indeed examples of the 10 percent of binaries that remain bound after a supernova explosion, then at least 10 percent of remnants that come from supernovae in binary systems might be expected to be related to such objects. Again, because of optical and X-ray obscuration, we could only expect to recognize such systems on the near side of the Galaxy. Nevertheless, the likelihood is that other such peculiar objects remain to be discovered.

If 10 percent of binary systems remain bound after one supernova explosion, about 10 percent of that number might be expected to remain bound after the later supernova explosion of the bound companion star. In other words, one in every hundred binary systems might eventually evolve to a final binary system containing two neutron stars—such is the nature of the binary pulsar.

Let us examine the light curves of supernovae to learn more about their nature. The early part of a Type II supernova light curve can be explained by the effects produced by the shock wave generated by the collapse of the core of the presupernova star. Energy is transferred to the star's outer layers, which are heated and accelerated. For some 30 days after the initial explosion, the visible surface of the star, its photosphere, expands at nearly constant velocity of the order of several thousand kilometers per second. The initial increase in the star's emitting surface results in an increase in radiated light. It must be stressed that supernovae are brilliant events not because they are hot but because they are large. However, continued expansion of the photosphere eventually results in a fall in temperature and thus brightness. By the time the photosphere has reached a radius of about 20 billion kilometers (greater than that of the solar system), the photosphere becomes so thin as to appear transparent. The star seems to shrink, although of course it is only the visible surface which is contracting. This should lead to a rapid decline in the light curve, as evidenced for the Type II supernovae.

The slower decay of the Type I light curves appears to require a continued input of energy. A possible postsupernova energy source previously mentioned was radioactive material (particularly radioactive nickel) generated in the extreme conditions of the supernova outburst. Appendix 17 discusses the nature of radioactivity and its role in the decay of Type I supernovae.

We can now concentrate on some of the remaining unanswered questions for the extended remnants of supernovae. A typical extended remnant can be simplified as having three components: first there is shock-heated interstellar material swept up by the expanding shock and expected to emit in X rays; then there is cooling, shock-heated material (supernova and presupernova ejecta, and interstellar material), which eventually cools to the point where it becomes visible at optical wavelengths; and finally there are the fast particles (cosmic rays) and compressed magnetic fields, evidenced as radio synchrotron emission. When observable, each component should tell us something slightly different.

A radio remnant obviously says very little about the nature of the supernova outburst itself. Possibly a plerionic radio remnant suggests that a Type II supernova produced an energetic pulsar that continues to illuminate the center of the remnant. The shell-type radio remnants tell us something about the weak interstellar magnetic field that permeates the Galaxy. The compression of this field is believed to create a magnetic field intense enough to produce, in association with the background cosmic rays, the detected synchrotron radiation. Most radio remnants are found to be concentrated to within a distance, referred to as the scale height, of about 250 light-years of the Galactic plane. This almost certainly does not mean that all supernovae occur that close to the plane. Young pulsars are found to lie usually within about 500 light-years of the plane, and these might be expected to be a better indication of how close to the plane supernovae themselves, or at any rate the Type II's that might produce observable pulsars, typically occur. Old pulsars have a much greater scale height of over 1,000 light-years, since their high velocities have carried them far from their point of birth.

It seems that the interstellar magnetic field that plays such a dominant role in forming the old radio supernova remnants is more closely confined to the plane than Galactic supernovae. Only supernovae close to the plane would be expected to appear as radio remnants, following the early phase when trapped magnetic field from the exploding star is used to explain the radio emission. If this argument is right, then radio remnants far from the Galactic plane should be seen as fainter on average than those close to the plane. This indeed seems to be the case for the majority of remnants; for example, the remnant of the supernova of 1006, which occurred over 600 light-years from the plane, is radio subluminous. One might also expect that large, old remnants should be brighter on their side closer to the plane, and this is also observed. It follows from this that usually it is only the supernovae that occur close to the plane of the Galaxy that will eventually produce long-lived radio remnants. This immediately explains the point made in Chapter 4, that long-lived radio remnants seem to have been formed, on average, about once every 80 years compared with the estimate for supernovae themselves of once every 20 to 30 years. The conclusion reached there was that only about one in every three or four supernovae occurs in regions that are suitable for the creation of long-lived radio remnants.

The material visible in optical supernova remnants is believed to be from three sources. The first is the interstellar medium, which is mainly hydrogen and helium but is seeded with debris from supernova explosions at earlier epochs. The second is circumstellar material—in other words, matter shed by the presupernova star during periods of instability in the later phases of its evolution. The final source of material is the ejecta from the supernova itself. Interpretation of the spectra of optical supernova remnants has revealed evidence of all three sources.

By interpreting the spectra of the optical supernova remnants, it is possible to gain information not only about the likely source of the emitting material (interstellar, circumstellar, or ejecta) but also about its temperature, density, and composition. Eventually, in association with other observational data, this can lead to an interpretation of the possible nature of the presupernova star. For example, the progenitor of Cassiopeia A has been modeled as a truly massive star, possibly as large as 100 solar masses. By contrast, the progenitor of the Crab may have been no more than 10 solar masses. Furthermore, by analyzing the composition of the material recognized as ejecta, it may be possible to reach an understanding of the nucleosynthesis of certain elements in the extreme conditions of supernova explosions. Finally, study of the interstellar material heated to millions of degrees by the expanding shock and emitting X rays provides details of both the temperature and the quantity of swept-up material. The amount of energy needed for the shock to sweep up this amount of material and heat it to the extreme temperatures inferred from the X-ray emission makes it possible to estimate the energy of the supernova explosion itself.

From the study of supernova remnants in our galaxy, as well as supernovae in external galaxies and historical records of supernovae, astronomers have been able to infer a great deal about the details of supernova explosions. By using such information, together with an improved theoretical understanding of how stars evolve, it has been possible to arrive at presently accepted ideas of what supernovae are, why certain stars undergo supernova explosions, where these occur, and how often.

Nearly all information we receive from supernovae and their remnants is in the form of radiation, although cosmic rays allow us to sample the actual material from the exploding superstars (see Appendix 18). But nearly all the material of our experience, including ourselves, had its ultimate origin in the stars. Scientists are now certain that all the material on Earth was created by a generation of stars that evolved before the birth of the Sun and planets. The role of the superstars in the creation of the elements and the solar system is the subject of the next chapter.

10

STARS OF CREATION

> There is nothing on Earth which is not in the heavens in a heavenly form, and nothing in the heavens which is not on the Earth in an earthly form.
>
> —*Emerson*

THERE ARE 92 known chemical elements that occur naturally. These range from the familiar (carbon, oxygen, silicon, and so on) to the rare (such as gold, silver, platinum), from the stable (iron) to the unstable and comparatively short-lived (uranium). But are the chemical elements that we find and study on planet Earth the same as exist elsewhere in the universe? Is the chemical composition of the universe stable or always changing? Has the medieval alchemist's goal of changing the elements from one form to another ever been achieved? And what part have the superstars played in creating the constituents of the universe as we now observe them?

The abundance of the elements can be studied both directly and indirectly. Direct sampling and chemical analysis has been possible for the Earth, the Moon (with the return of lunar samples from the manned Apollo missions), and for extraterrestrial debris reaching the Earth in the form of meteorites and cosmic rays. Remote sampling of Mars has also been achieved from the Viking Lander spacecraft, which sampled the surface and completed simple on-site tests of chemical composition and then telemetered the results to Earth. A continuing challenge for science in the third decade of space exploration will undoubtedly be remote investigation of the surfaces of other planets in the solar system and a rendezvous with a comet.

The alternative technique for studying the composition of astronomical bodies, and the only method available beyond the reach of spacecraft, is the use of spectroscopy. As described earlier, a line spectrum (either in emission or absorption) is a unique characteristic of an element. A complex spectrum can be solved to identify the many different elements contributing to it. In this way spectroscopy, over a range of wavelengths, has made it possible to infer the composition of the stars, nebulae, and interstellar medium. Helium, for instance, was found in the solar spectrum before its discovery on Earth. Spectroscopic studies reveal that, on a cosmic scale, hydrogen and helium are by far the most abundant elements, constituting about 98 percent, by mass, of the universe. The relative abundances of elements in the solar system have now been determined with some precision by solar spectroscopy supplemented by the direct measurements discussed above.

The creation of new elements in nuclear reactions is called nucleosynthesis. In the case of the simplest of the chemical elements, hydrogen, each atom is made up of just a single proton with its orbiting electron. The fusion of hydrogen nuclei at extreme temperatures to form helium, with the release of energy, was identified earlier as the principal energy source of a typical star for most of its life. This process of creating helium is usually called the proton-proton chain reaction. Alternatively, if carbon and nitrogen are already present, helium may be formed through the carbon-nitrogen cycle, the net result of which is, as for the proton-proton reaction, the transformation of four hydrogen nuclei into a helium nucleus.

After the long, stable hydrogen-burning phase of a star, the helium-burning phase that follows results in the fusion of three helium nuclei to form a carbon nucleus, containing six protons and six neutrons, with the subsequent reaction with a further helium nucleus producing oxygen. The product of fusing two helium nuclei, a form of beryllium, is unstable, and breaks up again into two helium nuclei. This quirk of nature has played a critical role in determining the composition of the universe. Scientists and science fiction writers alike have contemplated the effect on the universe, and indeed on life itself (since all organic compounds are derived from material containing carbon), if the form of beryllium containing four neutrons had been stable.

Heavier elements can be formed by the successive fusion of more and more helium nuclei, as occurs in later burning phases at increased temperature in the cores of massive stars. This production mechanism, with helium nuclei as the basic building block, is called the alpha process. In this way, for example, oxygen, neon, magnesium, silicon, and sulfur may be formed. It is believed that by subtle modifications of this process, elements with mass increasing to that of iron, and whose total numbers of protons and neutrons (a quantity known as the mass number) are multiples of four (the mass number of helium), can be created. To find elements of intermediate mass whose mass numbers are not multiples of four, one must look both at radioactivity and at the host of nuclear reactions that occur at temperatures above those necessary for the alpha process to proceed as far as iron. The extreme conditions of the supernovae of massive stars provide the required breeding ground for these so-called explosive nucleosynthesis reactions, which must take place without significantly destroying the major nuclei that have already been formed. Explosive nucleosynthesis puts the finishing touches to the mixture of lighter elements.

Element formation by successive buildup of nuclei of lighter elements, and related processes, can proceed only as far as the so-called iron-group elements—iron, cobalt, and nickel—since subsequent fusion requires the input of energy rather than the emission of energy of earlier fusion stages. Elements with mass number greater than for iron, which is 56, may be formed by the process of neutron capture. Neutrons, of course, carry no electric charge and therefore do not suffer electrical deflection by positively charged nuclei. They can therefore interact with the nuclei of heavy elements. The capture of neutrons can happen in various ways. In the so-called r process (*r* for "rapid"), several neutrons are captured by a single nucleus before it has time to decay by beta emission. By contrast, in the s process (*s* for "slow") an unstable nucleus, newly created by neutron capture, has time to decay before the arrival of the next neutron. The mixture of elements formed by the r and s processes differ. For example, gold and uranium are thought to be *r* process creations, while copper and lead are thought to be s-process creations. The r process produces the more neutron-rich isotopes; the s process bypasses these neutron-rich nuclei and produces many of the elements between the mass of iron and bismuth. There remain, however, certain proton-rich heavy elements that could not have been formed by either the r or s process. For these heavy elements, a rare capture event known as the p process (*p* for "proton") has been invoked, in which r- and s-process material is believed to be exposed to a fast flux of protons.

Table 3 summarizes the nuclear-fusion reactions that take place in stellar interiors and that are responsible for the creation of the elements lighter than iron. The s process takes place inside red giants, probably during the helium-burning stage of stellar evolution and probably in the region between hydrogen-burning and helium-burning shells. The red giants have convective outer layers that mix intermittently with the interiors, so products of nuclear processing may be dredged up to the surface layers where they can be observed spectroscopically. Anomalies in the abundances of carbon and nitrogen are observed and interpreted as evidence that a star has used some of its carbon to produce nitrogen in the carbon-nitrogen cycle which catalyzes the conversion of hydrogen to helium. Evidence for s-processed material is also seen; a dramatic example that reinforces the fact that we are witnessing recently synthesized material is the detection of the short-lived radioactive element technetium. Red giants release their outer layers either gradually, in the form of steady stellar winds or in a more sudden gust as a planetary nebula. This provides the mechanism for feeding the s-processed material and some alpha-pro-

TABLE 3

The Creation of the Elements

		Temperature (degrees)	*Mass*
Fusion reactions in stellar interiors	*proton-proton chain and CN cycles* (hydrogen to helium)	10 million	solar-type stars
	helium burning (helium to carbon, carbon plus helium to oxygen, oxygen plus helium to neon)	100 million	↓ increasing mass
	carbon burning (carbon to magnesium, sodium, aluminium, and neon)	500 million	
	oxygen burning (oxygen to silicon, sulfur, calcium)	1 billion	
	silicon burning (silicon to nickel, cobalt, and iron)	2 billion	red super-giants
Capture processes	*s process* (heavy elements created in neutron capture by iron group, plus beta decay)	in cores of red giants	
	r process, *p* process (heavy elements created by fast fluxes of neutrons and protons)	in supernova explosions	
Miscellaneous processes	*x process* (required to explain abundance anomalies, such as the light elements lithium, beryllium, and boron)	due to cosmic ray bombardment	

cessed material to the interstellar medium from whence they contribute to subsequent generations of stars. There is strong evidence that the material of the solar system has been subjected to the s process. Nuclei that easily capture neutrons are quickly destroyed when there are neutrons about, so it would be expected that the greater an element's ability to capture neutrons, the less there will be of that element. The quantity that is a measure of a nucleus's ability to capture neutrons is called its capture cross-section, which can be measured experimentally. The s process predicts that the product of the abundance of an element and the capture cross-section is constant over a range of atomic masses. The abundances of the elements in the solar system follow the s process prediction rather well (see Figure 33).

But what of the r and p processes, which require a fast flux of neutrons and protons? The extreme conditions required to initiate the r and p processes may occur in the exploding enve-

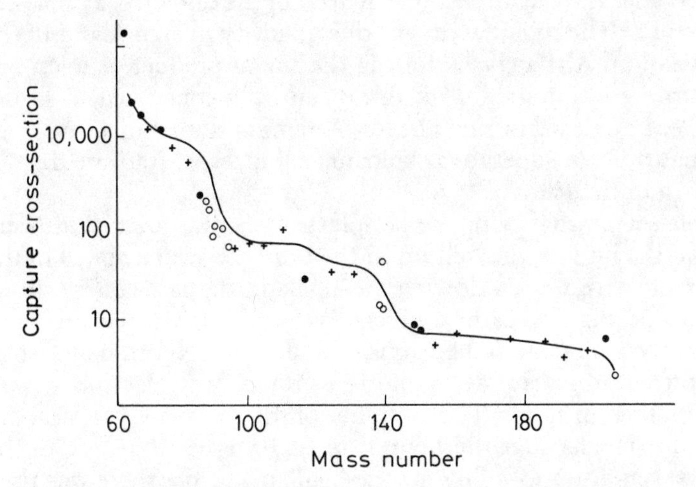

Figure 33. The s-process curve for solar-system abundances. Filled circles represent isotopes with known values; crosses, isotopes with estimated values; and open circles, isotopes produced partly by the r process, for which a contribution has been subtracted. The solid line is the theoretical result.

lopes of stars in supernova outbursts. The capture processes are also summarized in Table 3. It is interesting to note that the ultraheavy cosmic rays appear to be dominated by r-process rather than s-process material—compelling evidence that a significant proportion of the cosmic rays are born in supernovae.

Stellar nucleosynthesis, together with the r and p processes of supernova explosions, can account for the estimated cosmic abundances of nearly all the known elements. There remain a few anomalies. For example, the light elements lithium, beryllium, and boron are rapidly destroyed in stellar interiors. The fact that they nevertheless exist requires an extrastellar production mechanism. According to one explanation, heavy cosmic rays colliding with atoms of interstellar material split into lighter fragments (a process known as spallation), including lithium, beryllium, and boron. Such explanations required to explain abundance anomalies are known collectively as x processes.

It appears that almost one quarter of the universe, by mass, is helium. The manufacture of this quantity in stars is out of the question. Although helium is the major product of hydrogen burning in all stars, it is almost entirely consumed in subsequent nuclear-burning phases. Attempts to manufacture large quantities in supernova explosions of massive stars are fraught with difficulties.

It seems that both the simplest elemental form, hydrogen, and the bulk of the helium in the universe were created in the first few minutes following the Big Bang. In the seconds following the Big Bang, the universe was believed to be predominantly a mixture of neutrons and photons. Neutrons in isolation are unstable, decaying to a proton plus electron within about ten minutes. The existence of the observed cosmic abundance means that the helium must have been created in the first ten minutes following the Big Bang, before there was time for more than half the original neutrons to decay. In the primordial neutron-proton-electron soup, protons would have eventually captured electrons to form hydrogen atoms. Collisions of neutrons and protons would have produced nuclei of heavy hydrogen, called deuterium, and collisions of deuterium nuclei would have created helium. Attempts to re-create mathematically the conditions in the first few minutes following the Big Bang are able to explain satisfactorily the observed abundance of both helium and deuterium. The picture of element creation is now complete: hydrogen, deuterium, and helium were formed in the primordial universe, almost all elements as far as the iron group (and some beyond) came into existence in the interiors of stars, and selected rare heavy elements were created in the extreme conditions of supernova explosions.

While playing only a minor role in the creation of the elements, supernovae play a major role in their distribution. Though the bulk of the chemical elements are generated in the interiors of the stars, they are crucially dependent on supernova outbursts to free them into interstellar space. If the models proposed for Type I supernovae involving the creation of up to a solar mass of nickel decaying via cobalt to iron are correct, then Type I supernovae are probably the principal mode of seeding the interstellar medium with iron. Some alpha-processed elements and their derivatives will be fed through Type II supernovae. The very massive stars, greater than 25 solar masses, although relatively few in number, also have an important role to play. Evolving rapidly, they become bloated and unstable and undergo very significant mass loss; regions of nucleosynthesized material are peeled from such massive stars like layers from an onion. An extremely rare class of star known as the Wolf-Rayets (after their discoverers), extremely hot and massive, shows clear evidence of significant mass loss to reveal inner surface layers of processed material. Wolf-Rayet stars have been discovered at the center of about one fifth of known planetary nebulae.

From the interstellar medium, still predominantly primordial hydrogen and helium but seeded with heavier elements released by stellar winds, mass loss, and supernovae, new stars will be formed, some of them sufficiently massive to evolve eventually into supernovae. Subsequent evolution of the newly formed massive stars will further increase the concentration of the heavy elements and again enrich the interstellar medium after eventual supernova explosions. By this cycle of star formation, element creation in stellar interiors, and dispersal in supernova outbursts (see Figure 34), the concentrations of heavy elements in the interstellar medium will be gradually increased. It is fascinating to speculate that the bulk of the

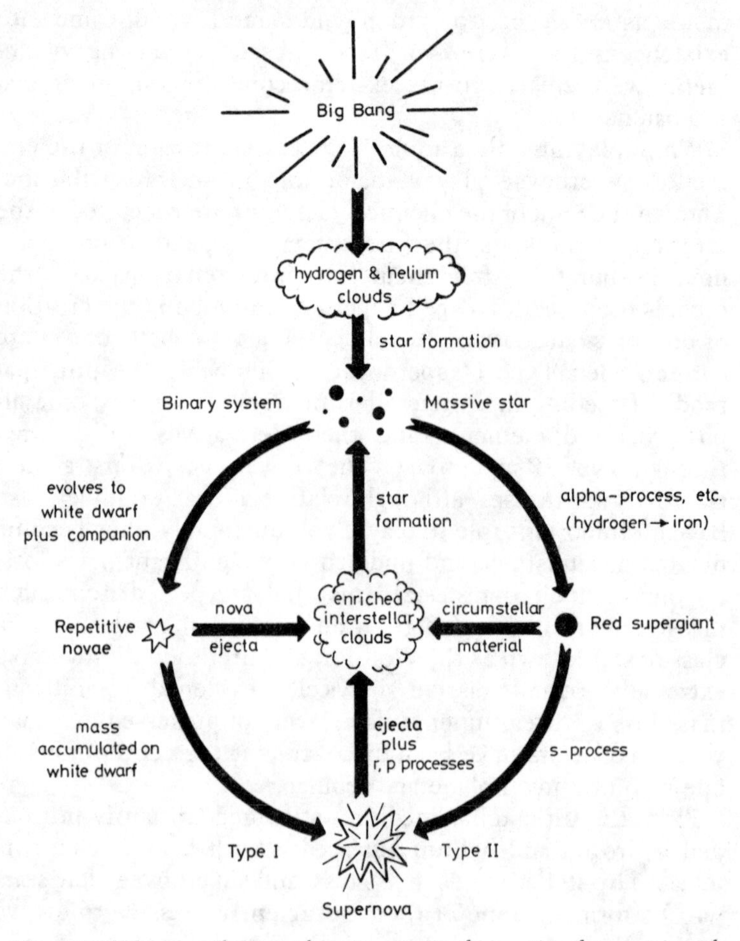

Figure 34. The contribution of supernovae to the continual processing of the elements.

familiar elements were created in the interiors of stars and reached us via supernova explosions—events that provided our very substance.

On the basis of the above arguments, one would expect elements created in stellar interiors by fusion reactions or by the s process to be more abundant than elements thought to be formed in the extreme conditions of supernova outbursts by the r and p processes. This is indeed the case, on both the terrestrial and cosmic scales. Thus iron, the end point of stellar fusion reactions, is more common than copper and lead formed by the s process in the center of evolved stars. These elements, in turn, are more common than platinum or gold, which are formed in the r process. Since supernovae are comparatively rare events, the precious r-process metals are also rare. Even rarer are the p-process elements, such little-known materials as molybdenum and samarium. The fact that the rare heavy elements are found at all testifies to the occurrence of supernovae throughout the history of our galaxy. In a galaxy with a much higher supernova rate and at a much more advanced stage of chemical evolution, gold, silver, and platinum, for example, would not have the same rarity value that they have on Earth.

There is circumstantial evidence that the supernova rate in the nascent Milky Way galaxy must have been very much greater than we observe now. Two distinct populations of stars are recognized in our galaxy. The first group consists of stars whose spectra show the presence of significant quantities of heavy elements; they are said to be metal rich. The stars of this group, the Population I objects, are comparatively young and were formed from interstellar clouds containing material processed in stellar events of an earlier epoch. The Population I stars predominantly lie close to the plane of the Galaxy, within the spiral arms. Old stars, formed shortly after the birth of the Galaxy and before the creation of heavy elements had proceeded far, appear metal poor in spectral studies. They belong to Population II. The fact that Population II stars show any evidence at all of processed material, however small a quantity, is difficult to explain if the primordial Galaxy was just hydrogen and helium. One possible explanation is that in the process of its formation the Milky Way first produced massive, short-lived stars, which rapidly evolved. Supernovae occurring as often as 10 per year may have fed heavy elements to the infant Galaxy, so that when less massive, long-lived stars began to form, they could contain at least small quantities of the heavy elements.

As already noted, the products of r-process and s-process syntheses are not identical, r-process products being neutron rich compared to s-process products. One can therefore determine the dominant route to the formation of the material in the solar system. As already noted, investigations suggest that from the abundance of heavy elements in terrestrial samples, meteorites, lunar samples, and so on, most must have been manufactured in the s process. Nevertheless, supernovae and the r process did play an important role, and certain r-process products have a built-in label of their date of manufacture. These are the radioactive elements, such as uranium. Two varieties or isotopes of uranium have mass numbers of 235 and 238 (each having 92 protons per nucleus, differing only in their number of neutrons). Both are products of the r process, being formed in about equal quantities. The 235 uranium decays by radioactive disintegration much more rapidly than the 238 variety. A measure of the rate of radioactive decay is a quantity called half-life, which is the time for half the number of atoms in a sample to undergo radioactive decay. The half-life of the 235 uranium isotope is about 700 million years, while the 238 isotope's half-life is six times longer.

Terrestrial samples of uranium show that the abundance of the heavier isotope now exceeds that of the lighter isotope by more than 100 times, even though they were formed in about equal quantities. From such information one can infer a date of manufacture for the bulk of the uranium in the solar system. This was over six billion years ago, more than one billion years before the formation of the solar system. This means that more than one billion years before the formation of the solar system, supernovae were manufacturing the uranium mankind now uses to fuel nuclear power stations. Even more remarkable, it is impossible to avoid the conclusion that the power of a supernova triggered the formation of the solar system itself.

Mention was made earlier of Laplace's celebrated nebular hypothesis of 1755. Laplace had envisaged that the solar system formed from a rotating interstellar cloud collapsing under the action of gravity. The mass of gas was molded to the shape of a flattened disk as the rotation rate increased with continuing collapse—again the analogy of a pirouetting ice skater may be used. Condensations in the disk then formed the planets, and the bulk of the gas remaining at the center eventually collapsed to form the Sun. Such a simplistic model, of course, leaves some questions unanswered. Studies showed that a rotating gaseous disk could not aggregate to distinct planets. Even if it could, the existence of nebular material out to the distance of the orbit of Neptune would mean that the eventual central collapsed object, the Sun, containing 99 percent of the mass of the original nebula, would have to rotate very much more rapidly than we see it rotating today.

Alternative theories were proposed. Could the collision of two stars have produced the fragmentation that eventually evolved to a planetary system? Less catastrophic, could the close encounter of the Sun with a passing massive star draw out a jet of solar material under gravitational interaction? The planets would then form from the material in this jet, being swept sideways into orbits about the Sun by the gravitational attraction of the receding star. But planets so formed could not rotate more rapidly than the parent Sun, as the majority of the planets do.

The underlying problem with all theories that claim that the material of the planets originated from the Sun itself, or from the presolar nebula, is the current orbiting speed of the planets. At such speeds, the preplanetary material at the surface of the Sun would have had so much energy as to be thrown off the Sun, escaping its gravitational field entirely; it could not have remained trapped in planetary orbits. Various solutions to this problem have been proposed. One suggestion is that before the existence of the planets, the Sun was part of a binary system. Material drawn from the Sun's binary partner in an encounter with a passing massive star could now orbit at the planet's observed speed. The encounter might also have dislocated the binary system and allowed the Sun's early companion to escape into space.

Present-day theories of the origin of the solar system return to the nebula hypothesis, though they do not necessarily involve a presolar disk, since the planetary material might be captured by the Sun during its passage through clouds of gas and dust in interstellar space. If this were the case, the majority of stars might be expected to acquire planetary systems. Attempting to account for the high orbital speeds of the planets compared with the present rotation rate for the Sun, theorists have proposed an intense magnetic field for the nascent Sun, coupling it to the condensations in the nebula forming planets.

One very important set of observations enabled scientists effectively to eliminate the models for the solar system that required the planets to have been formed from material extracted from the Sun (due to either a collision or a close encounter with another star). These have been the measurements of the ratio of deuterium to hydrogen in interstellar space, in the atmosphere of Jupiter, and in the solar photosphere. The deuterium to hydrogen ratio is about the same for the atmosphere of Jupiter and for interstellar space. In the case of the Sun, however, there is no evidence of deuterium at all, which is not surprising since deuterium is unstable under the extreme conditions found in stars. Thus Jupiter, and probably the other planets, must presumably have condensed directly from interstellar material, via either a nebula or direct capture by an already-formed Sun, without having gone through the intermediate stage of inclusion within the solar surface. The abundance of terrestrial lithium, beryllium, and boron, all unstable in stellar interiors also demands such a conclusion.

If we look again at the general question of how stars are formed, we can see that while it is accepted that stars are born from the collapse of interstellar clouds of gas and dust, an isolated cloud will not collapse without the assistance of some external compression force. As previously noted, in the disk of our own spiral galaxy, major episodes of star formation have been associated with compression lanes along the concave edges of the spiral arms. These compression lanes can be seen in photographs of external spiral galaxies as dark zones of dust bordering the spiral arms. The compression lanes are believed to delineate the density-wave pattern superimposed on the material of a galaxy, rather like the compressions and rarefactions of a sound wave in air, although just how the spiral density wave is generated or sustained remains poorly understood. As the material in the Galaxy rotates about the Galactic center, it encounters a compression lane that is associated with each of the two main spiral arms of the Galaxy. A cloud of interstellar material entering the leading edge of a spiral arm would be compressed. At the solar distance, a cloud would take about a million years to traverse a compression lane. On emerging from the compression lane, the cloud might be expected to re-expand, although without establishing its initial identity.

In some cases, however, compression of a cloud would be sufficiently great for the cloud's self-gravity to take control, with subsequent collapse and fragmentation, and the formation of a new cluster of stars. Some of these newly formed stars would be expected to be of comparatively low mass and consequently of low luminosity—but long-lived. Others would be many times more massive than our Sun, evolving over just a few million years. The extravagant consumption of nuclear fuel in these massive stars means that they will be superluminous up until the time they die in Type II supernova outbursts, their lives having been confined to the spiral arms in which they were born. Not for them the typically 100 million-year excursions into the void between the spiral arms. It is in fact the short-lived, superluminous stars that define most strikingly the actual spiral structure we see in photographs of so many external galaxies. Such a conclusion agrees with the fact that not a single Type II supernova has ever been observed in an elliptical galaxy. It seems that in the elliptical galaxies, massive progenitor stars, which characterize the spirals and produce the Type II supernovae, are now rarely formed.

But is star formation induced only by galactic compression lanes? Another possible source of compression is the shock wave generated by a supernova. Estimates suggest that the shock wave from a supernova would be capable of compressing a typical interstellar cloud by possibly a factor of 10 or more—certainly sufficient to initiate star formation (see Figure 35).

So can we see stars being formed around supernova remnants? There is one obvious problem. Once the collapse of an interstellar cloud is initiated, it will be probably at least several hundred thousand years before it can be seen as young stars. After such an interval, most extended supernova remnants have dispersed and are no longer easily identifiable. Thus, an extended remnant, and the young stars it is expected to spawn, would not normally be expected to be seen at the same time. But one partial ring of emission nebulosity, in the constellation Canis Major, has been proposed to be a very old supernova remnant—possibly about 800,000 years old. And along the edge of this remnant, concentrations of young, bright stars have been discovered. The association remains circumstantial, being consistent merely with the more or less simultaneous occurrence of the supernova and star formation. Identification of a causal link still requires further research. If supernovae do trigger star formation, how important might such a mechanism be when compared with formation in the spiral-arm compression lanes? The volume of space occupied by supernova remnants is roughly comparable with that occupied by groups of young stars within the Galaxy, and this suggests that a significant fraction of stars are formed in supernova events. A form of chain reaction might be envisaged, a supernova producing new stars, some of these eventually evolving to supernovae that produce new stars, this process repeating itself.

If star formation may be induced by two possible mechanisms, can we say whether our star, the Sun, was formed in a compression lane or by a supernova? Some clever scientific detective work in fact supports the latter. The evidence that the solar system was born shortly after a nearby supernova explosion comes from studies of the chemical elements in meteorites. Meteorites are believed to be debris from the protosolar nebula not assimilated into the planets during the formation of the solar system. Anomalies have been found in the

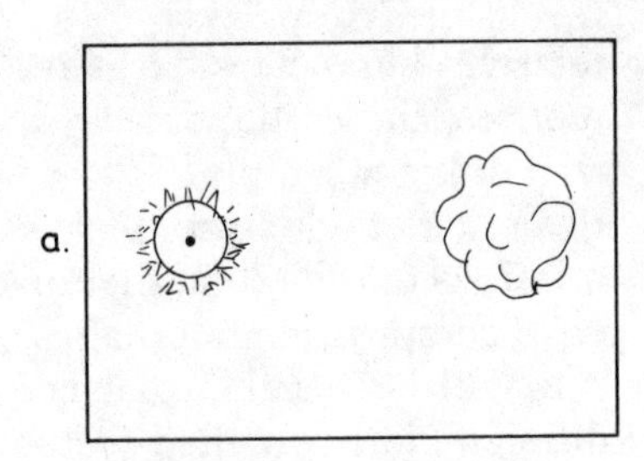

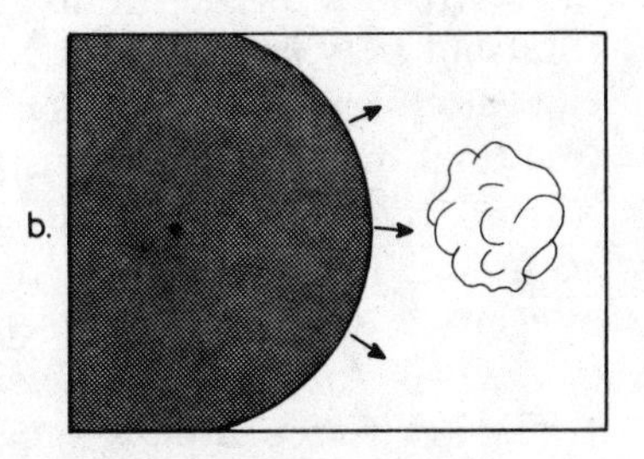

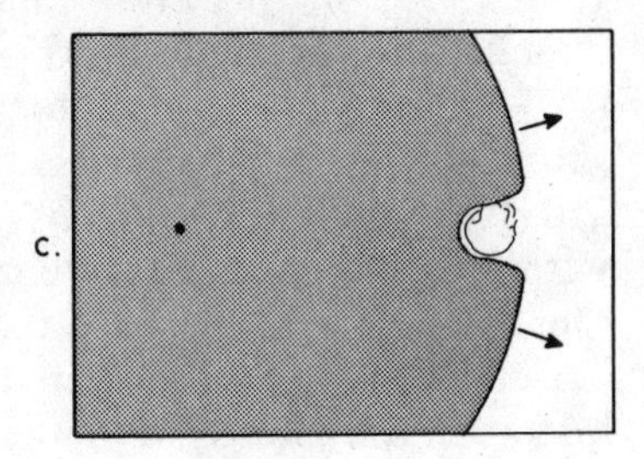

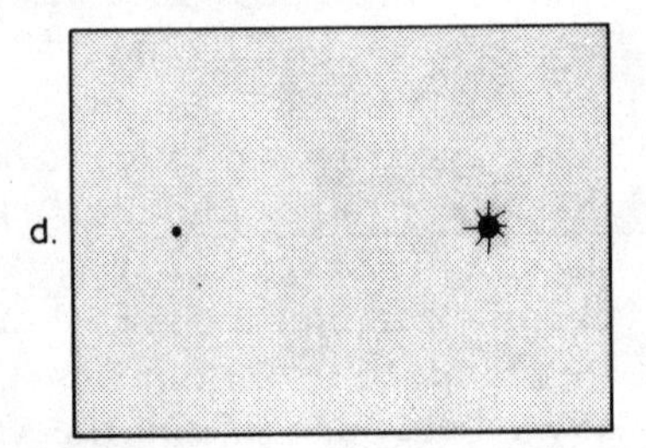

Figure 35. The supernova-induced formation of the solar system.

a. A nearby supernova.
b. The expanding supernova shock approaches the protosolar cloud—
c. Compresses it—
d. Forming the Sun and planetary system.

relative abundances of the isotopes of various elements in meteorites, these abundances differing significantly from terrestrial or lunar samples. It was in fact the sophisticated laboratories set up to analyze lunar samples from the manned Apollo missions to the Moon that made it possible to assess the meteorite composition anomalies with greater precision.

It will be recalled that certain chemical elements exist in different forms, called isotopes, in which individual atomic nuclei have the same number of protons but different numbers of neutrons. Oxygen atoms, for example, always have 8 protons, but may have 8, 9, or 10 neutrons in the different stable isotopic forms. These isotopes therefore have mass numbers (the total number of protons plus neutrons) of 16, 17, and 18. The relative abundances of these isotopes are constant to high accuracy. Terrestrial oxygen is known to be 99.756 percent oxygen-16, with 0.038 percent oxygen-17, and 0.205 percent oxygen-18. An identical division is found in lunar samples.

The key to recent advances in determining isotopic anomalies has been the analysis of samples from the fragments of a giant two-tonne meteorite that fell in 1969 near the village of Pueblito de Allende in northern Mexico. The Allende meteorite belongs to a class of meteorites known as the carbonaceous chondrites, so called because they contain carbon and are characterized by small inclusions or chondrules of molten material within them. Since the meteorites are believed to represent the most primitive form of solar-system material, which has been little altered since the meteorites solidified during the early history of the solar system, they are expected to bear witness to events associated with the creation of the solar system.

The Allende meteorite was to reveal several interesting isotopic anomalies. The first of these to be discovered was in the isotypes of oxygen, chemically isolated from the Allende material. Oxygen-16 was found to be relatively even more abundant than in terrestrial, lunar, or earlier meteorite samples, though the ratio of oxygen-17 to oxygen-18 was the same. The most obvious interpretation of this discrepancy was that the meteorite contained two components, the first having normal relative oxygen abundances and the second (representing up to 5 percent of the total oxygen content) being pure oxygen-16. The implication is that the protosolar cloud could not have been homogeneous. The normal oxygen content must have been thoroughly mixed, being uniformly maintained in the terrestrial and extraterrestrial samples, but the extra-pure oxygen-16 must have been present in some form in at least the region of formation of the Allende meteorite.

The implications of the next anomaly to be found in the Allende meteorite were to prove even more profound. This was the discovery of a decay product of radioactive aluminium-26. Stable aluminium is of mass number 27 (13 protons, 14 neutrons). Aluminium-26 decays by positive beta emission to magnesium-26, the half-life of the process being some 700,000 years. Magnesium-26 is just one of three stable isotopes of magnesium, the other two having mass numbers of 24 and 25. The relative abundances of the stable magnesium isotopes, like those of oxygen, are well determined from terrestrial and lunar samples. But again the Allende meteorite failed to conform to cosmic convention. It had a significant excess of magnesium-26, and it was demonstrated that this could only have originated *in situ* from the radioactive decay of aluminium-26.

Because aluminium-26 is radioactive, it must have been incorporated soon after its creation into the material of the meteorite, and certainly within a few half-live intervals, say about two million years. Aluminium-26 is formed in supernova explosions by the p process, when magnesium-25 (itself formed in the r process) absorbs a proton. It is unlikely that the aluminium-26 that led to the magnesium-26 anomaly in the Allende meteorite was produced by processes other than supernova synthesis, since alternative processes suggested predict other isotopic anomalies that are not observed. There seems no escaping the conclusion that a nearby supernova occurred shortly before the formation of the solar system. Aluminium-26 created in the supernova outburst was assimilated in some form into the material eventually solidifying as the Allende meteorite, there decaying to produce the magnesium-26 abundance anomaly identified almost five billion years later. The unrelated chance occurrence of a supernova just before the formation of the solar system might be considered possible, but it is much more likely that the supernova explosion that created the aluminium-26 also precipitated the collapse of the protosolar cloud.

A further meteorite abundance anomaly may reveal details of the solar-system prehistory. Traces of various isotopes of the element xenon, showing an unusual relative abundance, indicate their likely origin in the radioactive decay of iodine-129 and the spontaneous division (fission) of plutonium-244 nuclei. Both iodine and plutonium are created only in the r process, in the extreme conditions of supernova outbursts. In this case, it was possible to estimate that they were formed about 100 million years before their decay products were assimilated into meteorites. We therefore have evidence of another supernova occurring close to the protosolar cloud a full 100 million years before the birth of the solar system.

These pieces of evidence have enabled the following prehistory of the solar system to be proposed. The protosolar cloud, like other clouds of interstellar material, revolved around the Galactic center, passing alternately through the spiral arms, within which Type II supernovae tend to occur. The probability of an encounter between the cloud and a supernova within about 30 light-years on each transit of a spiral arm may be estimated to be about 50 percent. The transit time through the interarm regions at the solar distance is about 100 million years, just the interval estimated between the creation of the iodine-129 plus plutonium-244 and of the aluminium-26, the decay products of which appear in the meteorites. The addition of iodine-129 plus plutonium-244 to the protosolar cloud probably resulted from its last encounter with a nearby supernova when crossing a spiral arm. When next the cloud entered a spiral arm and encountered a supernova, which injected aluminium-26 into the cloud, the encounter was close enough to bring about the collapse of the cloud and trigger the formation of the solar system.

Much of the material of the solar system was created inside the stars and recycled to the interstellar medium in supernova explosions. An interstellar cloud, which contained material processed by supernovae, then collapsed under the influence of the expanding shock wave from a nearby supernova, giving birth to the solar system. In its subsequent history, the solar system must have encountered a nearby supernova at least once per orbit of the Galaxy. The superstars have therefore acted as procreator, midwife, and nursemaid to the evolving solar system.

SUPERSTARS AND THE DESTINY OF THE UNIVERSE

THE STORY of the superstars is one of their continuing influence on the destiny of the universe. In the cycle of star birth, life, and death, all the elements apart from hydrogen and helium have been formed, either from energy-generating reactions or in a star's final death throes. Matter is then blasted into space in supernova outbursts, which help to shape a new generation of stars—dust to dust, ashes to ashes.

It is a somewhat sobering thought that the carbon, nitrogen, and oxygen intrinsic to the organic compounds were formed in the centers of stars. Ancient mythologies relating human beings to the stars therefore contained a semblance of truth—we are indeed the children of the stars. Almost every single atom in our bodies has been processed through one or more supernova explosions at some time during aeons past. In its very moment of self-destruction in a supernova event, a star may procreate a new stellar system; in this way, it is believed, the birth of our own solar system was induced. Subsequently, supernovae may occur often enough within the neighborhood of a planetary system to influence profoundly any life-forms evolving there, either through direct energy input or by way of forced climatic changes. As the universe expands toward its still-debated goal, exploding superstars continue to stir the celestial soup.

At what rate do supernovae occur throughout the universe? To answer this question, a boundary for observations must be established, since in looking out into the cosmos we are also looking back in time. At extreme distance, the galaxies seen are in the making, so their supernova rates would not necessarily be characteristic of evolved galaxies such as the Milky Way. It is therefore helpful to set an arbitrary limiting distance of eight billion light-years. At such a distance, most of the galaxies are in a relatively advanced stage of evolution: the spirals would have settled into the routine processing and enrichment of the interstellar material as a result of massive star formation in spiral arms and subsequent supernova explosions.

Four hundred and fifty supernovae have been discovered in other galaxies over the past 90 years. These represent only a small fraction of those that have occurred during this interval, since the ground-based supernova surveys have been necessarily restricted in space and time. Sufficient numbers have been discovered, however, to derive statistically the supernova rates characteristic of different galaxy types—three or four per century in a spiral galaxy like our own, with lower rates in galaxies of different types. An average rate over all galaxy types of about one supernova per galaxy every 200 years seems a realistic estimate.

The spatial density of galaxies is well established from measuring the red shift of selections of galaxies and determining their distances by the application of the Hubble relationship. The universe seen at extreme distances is at an earlier stage of evolution, when the galaxies were more closely packed than the more recent universe seen nearby, although galaxies tend to cluster as the universe expands and evolves. Within the portion of universe of present interest, to a distance of eight billion light-years, the number of galaxies can be estimated to be of the order of 50 billion. Each one of these galaxies would have an average population of at least a similar number of stars. On this scale, planet Earth shrinks into insignificance. A supernova every 200 years on average in each of 50 billion galaxies means that supernovae are occurring within the studied portion of universe at the rate of about ten every second!

Just half a century ago white dwarfs were held to be the ultimate state of stellar evolution. Neutron stars and black holes existed only in the realms of science fiction. But a universe evolving without the superstars would have been stagnant indeed. It would have been made up almost entirely of hydrogen and helium created in the Big Bang, the bulk of processed heavier elements remaining trapped inside dead stellar remnants. Limited quantities of certain light elements may have been fed to space by stars shedding their outer layers, but a universe without the neutron and proton capture processes of supernova explosions would have been devoid of many elements heavier than iron. Ours would then have been a planet without the precious metals gold, silver, platinum—or without the radioactive elements such as uranium that have played so important a role in the geological evolution of the Earth. One of the main energy inputs to the galaxies would have been lost, since cosmic rays represent a heating source for the interstellar medium comparable to starlight. The fact that we now explore such fascinating objects as pulsars, black holes, X-ray binary systems, exploding stars, spectacular nebulosities, extended X-ray and radio objects, and runaway stars testifies to the fact that the superstars must be among the most important and spectacular phenomena in our evolving universe.

APPENDIXES

APPENDIX 1: Atoms

Atoms are the smallest recognizable components of the 92 known naturally occurring elements. Atoms themselves are composed of protons and neutrons, closely bound within a central nucleus, with minute electrons, a mere two thousandth of the mass of the protons or neutrons, orbiting the central nucleus. Electrons carry negative electrical charge, and protons positive electrical charge; neutrons carry no electrical charge. Electrons are constrained to well-defined orbits of particular energy, although transitions between orbits are possible. For an electron to be excited from one orbit to another at a higher energy level, energy must be supplied. If sufficient energy is supplied, the electron can be removed entirely—the atom is ionized. However, if an electron falls from a state of high energy to an available state at lower energy, the difference energy is given out in the form of a quantum of radiation—a photon. The relationships between allowed electron energy levels in an atom is a complex subject that will not concern us here, except to note the famous exclusion principle expounded by Wolfgang Pauli to describe one of nature's constraints on the behavior of matter—namely, that only two electrons in an atom can occupy the same energy state, and these electrons must differ in their spin. Electrons have spin since they behave in some ways like miniature gyroscopes; they have two possible ways of spinning, which can be thought of as clockwise and counterclockwise. Electrons with opposite spin form stable pairs, which can occupy a single energy level of an atom. Thus, the Pauli exclusion principle can be seen to have important ramifications for the structure of atoms. We shall see that it also determines the fate of the stars.

The number of protons within an atomic nucleus identifies an atom as that of a particular element. Thus, for example, the simplest elements are hydrogen and helium, with just one and two protons respectively. Carbon atoms have 6 protons, oxygen 8, iron 26, copper 29, gold 79, lead 82, uranium 92, and so on. In their so-called neutral state, atoms have as many electrons orbiting their central nuclei as protons within them. The number of neutrons in an atom is usually about equal to or somewhat greater than the number of protons. Uranium for example, has 146 neutrons tightly packed with the 92 protons in its nucleus; at the other end of the scale, normal hydrogen has just a single proton with no neutron, and helium has 2 neutrons accompanying the 2 protons. Certain varieties of the same element, called isotopes, retain the same number of protons although they have differing numbers of neutrons. Protons, neutrons, and electrons are the subatomic particles that will figure most prominently in our deliberations, but we will come across others.

APPENDIX 2: Energy from the Nucleus

Energy can be extracted from the nucleus of an atom by two important processes. In the first, a massive nucleus, of, for instance, uranium, splits to produce lighter fragments and the release of energy. This is the process of nuclear fission, utilized in atomic power stations and more catastrophically at Hiroshima. For nuclei lighter than that of iron (with 26 protons), energy is released when light nuclei are forged together to form a larger nucleus, again with the release of energy. This is nuclear fusion. In newly forming stars, the initially important nuclear fusion reaction is referred to as the proton-proton chain reaction converting hydrogen to helium. The stages of the proton-proton chain reaction are as follows: First two hydrogen nuclei (single protons) merge to form a heavy hydrogen nucleus of a proton plus neutron. (In this process, a proton is converted to a neutron, with the release of two subatomic particles—a positron of identical mass to the electron but opposite electrical charge and a strange, will-o'-the-wisp particle called a neutrino.) A heavy hydrogen nucleus then merges with a normal hydrogen nucleus to form light helium (two protons plus one neutron). Finally, two light helium nuclei merge to form a single normal helium nucleus (two protons plus two neutrons), leaving two residual hydrogen nuclei. The net effect is four hydrogen nuclei fusing to form a helium nucleus. In the next stage of stellar evolution, the helium ash left from hydrogen burning is converted to helium by the triple helium reaction. Mass is destroyed in these transmutations, and energy is released. The relationship between destroyed mass m and energy E is described by Einstein's equation $E = mc^2$, perhaps the most famous formula in all of physics. Because c is the velocity of light, c^2 is an extremely large number. This formula shows that the destruction of a small amount of mass results in the release of a vast amount of energy.

APPENDIX 3: Electrons in Stellar Interiors

Although the fast-moving electrons in the center of a star lack the order they are forced to observe when trapped in atomic orbits, their complicated trajectories do obey the Pauli exclusion principle. The number of possible electron trajectories in a volume of gas is very large but nevertheless finite. The Pauli principle then dictates that no more than two electrons, with the same energy but differing spin, can travel along any given "permitted" trajectory. In the hot gas of a normal star, the number of available electron trajectories is so great as to place no significant constraints on the electrons. This is not the case in later stages of the star's evolution, as, for example, when a solar-type star collapses to become a white dwarf.

APPENDIX 4: What Stops a White Dwarf from Shrinking Further?

When a star shrinks toward white-dwarf densities, the stellar core resembles a gaseous sea of electrons with embedded nuclei. Compression has reduced the number of permitted electron trajectories to a condition known as degeneracy, where all are occupied by rapidly moving electrons. Any further compression would decrease the number of available states to below the population of electrons—an impossible situation in the atomic world. The reluctance to assume a more compact state is described in terms of a degeneracy pressure exerted by the electrons. Thus, if the total mass of stellar material to be supported against self-gravity is not greater than the Chandrasekhar limit, 1.4 solar masses, electron-degeneracy pressure can sustain the star's structure at the white-dwarf stage.

APPENDIX 5: What Stops a Neutron Star from Shrinking Further?

At the extreme temperatures and pressure generated as the core of a massive star collapses, individual protons within the various nuclei and electrons merge. The fruits of this union are neutrons; the capture of an electron by a proton is sometimes referred to as neutronization. In the neutron building process, neutrinos are created, but most of these are trapped within the collapsing core. The loss of electrons reduces remaining electron-degeneracy pressure, further accelerating the collapse. Neutrons, like electrons, are constrained by the Pauli exclusion principle. Thus, it is possible for the degeneracy pressure exerted by the tightly packed neutrons to take control and stop the core from contracting further.

APPENDIX 6: Gravity Triumphs!

When neutron-degeneracy pressure is insufficient to halt contraction of a collapsing stellar core, there is nothing to save very massive stars. Gravity finally triumphs, driving the core of a giant star beyond nuclear densities to complete collapse as a black hole. The idea of a black hole can be explained in part by a simple thought experiment. When a ball is thrown in the air, it falls back to Earth under the action of the Earth's gravitational field. In the weaker gravitational field of the Moon, there would be no difficulty about throwing the ball very much higher. If only the ball could be thrown fast enough, it might escape the gravitational field completely and fly off into space. The minimum velocity needed to escape an astronomical body's gravitational field is known as its escape velocity. The escape velocity for the Earth's surface is about 40,000 kilometers per hour. The more intense the gravitational field of an astronomical body, the greater its escape velocity. A white dwarf would have an escape velocity of several million kilometers per hour. If the escape velocity of a body exceeded the velocity of light (300,000 kilometers per second), then even light itself could not emerge from the body. Such is the nature of a black hole, an object invisible since its intense gravitational field does not allow light to escape.

The maximum mass of a neutron star is believed to be about three times the mass of the Sun. If it is more massive than this, the degeneracy pressure of the tightly packed neutrons cannot balance the intense gravitational forces. A neutron star of one solar mass would have a radius of just 10 kilometers. If such a neutron star could be further compressed to a sphere with a radius of three kilometers, it would become a black hole. If the mass of the Earth were concentrated to become a black hole, it would be just a centimeter across. Since according to Einstein's

Theory of Special Relativity nothing can travel faster than the speed of light, material cannot escape a black hole. Its intense gravitational field will continue to suck in material but will never regurgitate it.

APPENDIX 7: The Wave Nature of Light

The experiments that revealed light to be a form of wave date from the seventeenth and eighteenth centuries. Light was found to display certain properties, notably interference and diffraction, normally associated with waves. The properties of common examples of waves are well known. A traveling wave is characterized by the distance between adjacent crests (the wavelength), the amplitude of the wave, and by the number of crests passing a stationary observer per second (the frequency), related in such a way that the wavelength multiplied by the frequency gives the speed of the wave. The wavelength of a particular light wave determines its color.

It was a Scot, James Clerk Maxwell, who recognized in 1865 that light waves were electromagnetic in nature. An electric charge exerts an electric force on any nearby electric charge, the mode of such interaction being assigned to an electric field generated by the charge. Similarly, a magnet via its magnetic field influences some other nearby magnet. Maxwell showed that any disturbance in an electric field would be expected to propagate outward as an electric wave, just as the disturbance of water in a pond generates a water wave. But the electric wave is always accompanied by a magnetic wave forming a composite electromagnetic wave. Maxwell's early theoretical considerations showed that, if electromagnetic waves existed, they would travel through free space with a speed of 300,000 kilometers per second, close to the best determined value for the speed of light available at the time. On the basis of this agreement, Maxwell suggested that light was a form of electromagnetic wave. Modern studies show that light is not emitted as a continuous wave but rather in small energy packets called

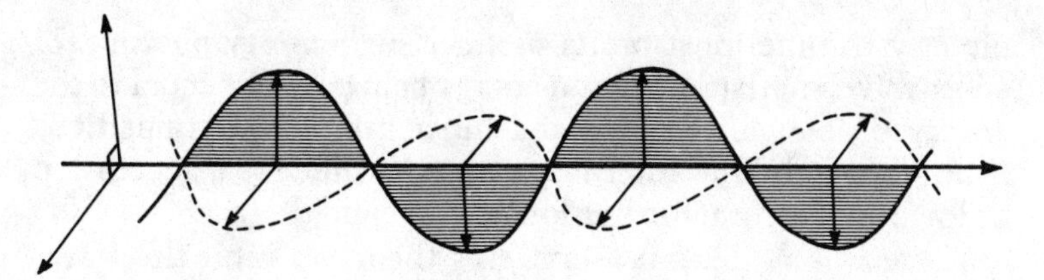

Figure 36. Schematic representation of an electromagnetic wave, made up of an electric wave (shaded) accompanied by a magnetic wave (broken line) such that the electric and magnetic components are always at right angles.

photons. A stream of photons, however, can be adequately modeled as a continuous electromagnetic wave. Electromagnetic waves are depicted in Figure 36.

The energy of visible-light photons is determined by the characteristics of the atoms from whence they come. As previously noted, electrons may move only between discrete orbits of different energy in an atom; inner orbits having lower energy than outer orbits. Lower-energy inner orbits are occupied preferentially. An activated or excited electron, given an increase of energy, may jump directly from an inner orbit to an available outer one but may then return to its original inner orbit via discrete steps to orbits of intermediate energy rather than in one leap (see Figure 37). Thus, transitions from outer orbits

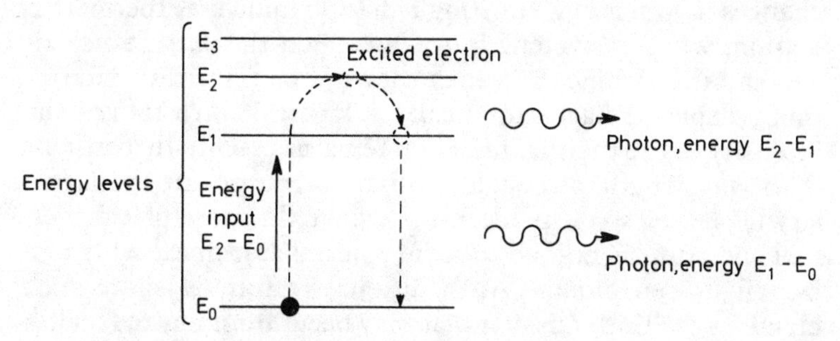

Figure 37. An electron excited to a higher energy level returns to its original state, emitting energy in the form of photons.

directly to innermost orbits produce high-energy photons, of short wavelength (the photon energy being exactly equal to the energy difference of outer and inner orbits); but transitions from outer orbits to intermediate orbits produce a succession of lower-energy photons of longer wavelength.

There was nothing in Maxwell's theory to limit the wavelength of electromagnetic waves, and we now recognize that visible light comprises just a small portion of the full electromagnetic spectrum.

APPENDIX 8: Thermal Radio Emission

The interstellar hydrogen in the Galaxy tends to be distributed in clouds in which the hydrogen normally remains in its neutral state. If, however, a hot star lies near or within such a cloud, ultraviolet radiation from the source will ionize the cloud by stripping the electrons from the atoms to leave isolated protons.

When an electron and proton in random thermal motion approach each other, they interact electrically. The interaction is termed free-free, since the particles are unbound before and after the interaction. The proton shows only a small deflection from its original path, but the lighter electron completely changes its direction, emitting radiation known as thermal radiation, whose wavelength depends upon the particle velocities and the distance at which they interact. Such an interaction is depicted diagrammatically in Figure 38. In a hot gas, the intensity of the thermal radiation remains essentially constant with wavelength, although beyond a certain point the intensity decreases with increasing wavelength. This typical thermal spectrum is unique to radio sources associated with ionized hydrogen clouds. An additional feature of some such clouds is the line emission that may occur from ionized hydrogen regions during a brief recombination of an ion and electron.

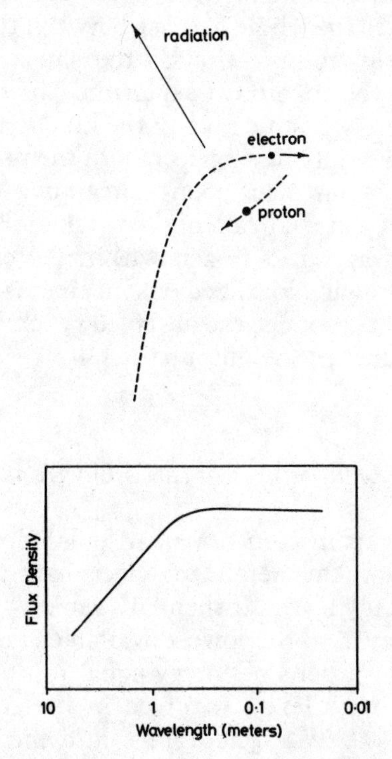

Figure 38. Thermal radiation produced when charged particles interact.

APPENDIX 9: Synchrotron Radiation in Supernova Remnants

Although the spectrum and polarization of the radio emission from supernova remnants confirm that it is synchrotron radiation, there is as yet no single completely satisfactory theory for the origin and evolution of the required high-energy particles and the magnetic fields. There is increasing evidence that these may be dominated by distinct physical processes at various stages of a remnant's evolution. For old supernova remnants (older than a few thousand years), the magnetic field required is believed to be just the weak interstellar magnetic field that pervades the whole of space between the stars but has been enhanced by compression as the shock wave expands from the site of the supernova explosion. The energetic particles are then merely cosmic rays, of the kind that permeate the whole Galaxy, possibly having been born themselves in earlier supernova explosions. For young supernova remnants, the magnetic field is just a portion of the magnetic field of the star that exploded, trapped or frozen within the ejecta from the explosion and showing localized enhancements. The energetic particles needed to produce the observed synchrotron emission are those generated in the outburst itself.

APPENDIX 10: Light from Captured Electrons

Although electrons in atoms of a particular element can have only certain allowable energies, an electron can be captured at any of these energy levels. It then falls rapidly, via a variety of intermediate states, to the lowest available energy level, with the emission of photons of energy equal to the energy differences of the atomic levels involved in the transitions. The transition from the hydrogen atom's third energy level to the second energy level produces the most intense visible radiation, at a wavelength of 6,563 angstroms—the so-called H-alpha emission. (An angstrom is one hundred-millionth of a centimeter.)

APPENDIX 11: Spectroscopy

The phenomenon of colors, in which a beam of white light passing through a prism produces the colors of the rainbow, had been demonstrated since ancient times. It was in the seventeenth century, however, that Isaac Newton, by a series of carefully thought-out experiments using the effect, and by logi-

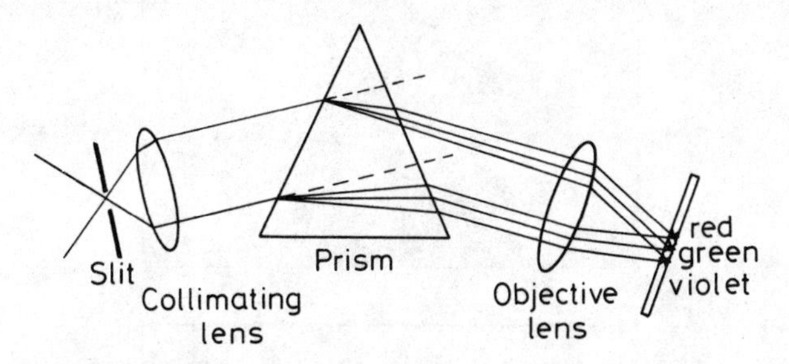

Figure 39. When a beam of visible light passes through a prism, its component colors spread into a spectrum.

cal reasoning, established that white light "is a confused aggregate of Rays indued with all sorts of colors." When a prism refracts or bends the different color components of white light by varying amounts (red being bent least, violet most), it separates a beam of white light into a merging row of colors—a spectrum (see Figure 39). Since white light, passing through a slit and incident on a prism, produces a continuous range of colors, it is said to have a continuous spectrum. To take a contrasting example, a certain purple light source might have only red and blue components, so that its spectrum would show just two features on a black background, making a so-called line spectrum.

Different elements emit light of different colors characteristic of electron transitions between the various energy levels of atoms of the element. The different colors can be distinguished as emission lines in a spectrum if the element is in gaseous form of low density and energy is supplied. An emission-line spectrum is a unique characteristic of an element, a fingerprint or signature that allows immediate and unambiguous identification (see Figure 40). A light source—for example, a supernova—may contain many different elements, and its spectrum must be carefully compared with emission-line spectra obtained in the laboratory for all the chemical elements to identify those making up the gas. If a beam of white light, displaying a continuous spectrum, shines on a low-density gas that

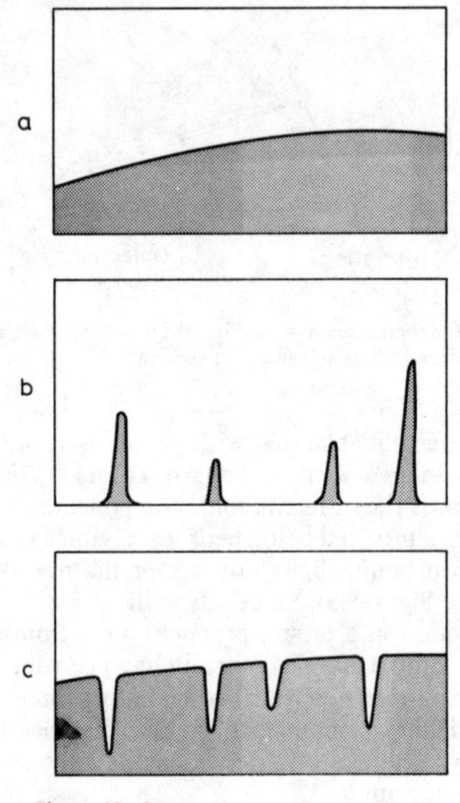

Figure 40. Spectra.
a. Continuous, showing a smooth distribution of light.
b. Emission line, with light emitted only in certain characteristic colors.
c. Absorption line, with light diminished at certain colors.

does not itself radiate, certain colors of the incident light are absorbed so that the emergent light shows dark lines in its spectrum against a continuous background. Such a spectrum is referred to as an absorption-line spectrum. The colors of light absorbed are the same as those that would be emitted by the low-density gas if it were made to radiate, so that an absorption-line spectrum of a gas reveals the same information as would its emission-line spectrum. Since the spectrum of a particular light source provides important information on the composition, density, temperature, and velocity of the light source, spectroscopy—the study of spectra of different light sources—has proved to be an extremely powerful observational technique in astronomy.

APPENDIX 12: The Doppler Effect

Wavelength shift is an effect common to any wave motion. First formulated mathematically by Christian Doppler in 1842, the phenomenon now bears his name. A familiar example of the Doppler effect is the sudden decrease in pitch of the siren of a passing police car noted by a stationary observer. Sound, like light, is a wave. An approaching sound source compresses the wavelength of the emitted sound in the direction of motion, without affecting the sound velocity, which depends only on the state of the intervening air. If the velocity is unchanged, a decrease in wavelength implies a corresponding increase in frequency (pitch). Similarly, when the sound source recedes, it stretches out the wavelength so that the pitch sounds low to the stationary observer. And so it is with light. The compression of wavelength from an approaching light source produces a blue shift, whereas the red shift is interpreted in terms of motion away from an observer. The degree of blue shift or red shift allows an estimate to be made of the speed of approach or recession of the light source.

APPENDIX 13: Measuring Distances to Galaxies with Supernovae

Supernovae are in a unique position to help finally establish the true distance scale of the universe. Accurate spectroscopic observations of a supernova and estimates of its brightness give both the temperature of the supernova's emitting photosphere and the amount of energy received from it at the Earth's surface. The amount of energy emitted from the surface of any hot body of known temperature depends only on the area of its surface—that is, on the square of its radius. The fraction of this energy received at a remote point then varies as the inverse square of the distance. This means that measurements of a supernova's temperature and the energy received from it at the Earth give the ratio of radius of the supernova to its distance. Observations taken over a period of weeks allow the increase in radius of its photosphere with time to be estimated, while Doppler shift measurements will show how fast the supernova is expanding in absolute terms. This information then allows the distance to the supernova and its parent galaxy to be estimated free from many of the ambiguities and uncertainties that have plagued other methods. The technique is still in its infancy, but it is possible that the superstars may be the key to estimating the true size and age of the universe.

APPENDIX 14: 21-Centimeter Emission from Hydrogen

The 21-centimeter emission is associated with the spin state of the single electron in the lowest energy level of neutral hydrogen. The energy of an electron differs very slightly between the two available spin states—in one case the spin of the electron is in the same direction as the spin of the nucleus proton and in the other case it is in the opposite direction. A hydrogen electron in the higher-energy spin state can flip to the other allowable spin state spontaneously, with the emission of the characteristic 21-centimeter radiation. Although the probability of such a flip is very low, the enormous number of hydrogen atoms in a significant volume of the Galaxy results in a detectable hydrogen line. The wavelength of the hydrogen emission detected from any direction in space will not be concentrated at 21 centimeters, but may show considerable spread in wavelength. There are two reasons for this, both involving the Doppler effect described in Appendix 12. The first is that in any volume of space containing a cloud of hydrogen, individual atoms will be in random motion. Those moving away at time of electron spin flip will show a red shift (wavelength increase) and those toward the observer a blue shift (wavelength decrease), so that the line detected from a particular hydrogen cloud will be broadened, reflecting the random motions of the emitting atoms. The second reason is given in terms of large-scale motions within the Galaxy. Measurements of nearby stars show that the disk of the Galaxy is rotating, not as a solid body but differentially—the farther one is from the center of the Galaxy, the longer the time interval for a single circuit. As previously mentioned, at the Sun's distance the time for one revolution is estimated to be about 200 million years! If the interstellar hydrogen is moving with the stars, then along any line of sight we might expect to see clouds moving with different velocities with respect to the solar system and showing varying Doppler shifts. The resulting profile of hydrogen 21-centimeter emission will thus be extremely complex, displaying many components associated with emission from clouds at varying distances and with varying velocities relative to the solar system, extending to the outer limits of the Galaxy, since the radio waves suffer negligible attenuation on their long journey. From models of rotation of the Milky Way, a particular peak in a complex 21-centimeter profile can be associated with emission originating in a hydrogen cloud at a certain distance. Using this technique, the neutral hydrogen in the Galaxy has been mapped and displayed as if viewed from above the plane of the Galaxy. Such observations confirm the complex spiral structure of our galaxy and identify the individual spiral arms.

Neutral hydrogen can also be seen in absorption. If a source of continuum radio emission, such as a radio supernova remnant, is viewed through a hydrogen cloud, the 21-centimeter feature will be seen as a dip in the spectrum of the continuum source. The broadening and wavelength displacement of the dip will reflect the properties of the intervening hydrogen cloud. In particular, the Doppler shift of the dip can give the distance to the intervening cloud (or clouds if there are several absorption dips). This technique can give a lower limit to the distance to the source of continuum radio emission—it must lie beyond the most distant intervening cloud. This technique has been applied to give minimum distance estimates to many Galactic supernova remnants.

APPENDIX 15: The Dynamical Evolution of Supernova Remnants

In recent years, very detailed mathematical models involving the use of giant computers have provided a fairly detailed picture of how a remnant evolves with time. Despite the complexity of these models, a simplified picture of remnant evolution may be formed by considering the process in four parts or phases.

In Phase 1—the free expansion phase—the effect of the interstellar medium is ignored, and the expansion of the ejecta from the supernova explosion may be considered to be essentially free and proceeding at constant speed. The free-expansion phase lasts for just a few hundred years, after which the decelerating effect of the interstellar material swept up by the expanding shock wave from the initial supernova explosion begins to be felt. The expanding supernova remnant has now entered Phase 2.

Phase 2 is the so-called adiabatic expansion phase, *adiabatic* being the scientific term applied to any system which is neither receiving nor losing energy. In this particular case, the strict application of the term is incorrect, since the expanding remnant may continue to receive an injection of energy from an active central pulsar. Furthermore, the remnant is certainly losing energy through radiation over a range of wavelengths. In a simplified approach, the possible injection of pulsar energy is not considered; furthermore, since for many thousands of years the radiated energy is such a small fraction of the total energy of the remnant, it too is ignored. The adiabatic assumption is therefore an acceptable approximation to reality for middle-aged remnants. The adiabatic-expansion-phase model was originally investigated in studies relating to the behavior of the expanding shock wave generated by atomic bombs. Atmospheric atomic bomb tests were observed so that the theory could be improved on the basis of experimental observation. With the realization that supernova explosions represented a similar phenomenon on a much grander scale, theoretical astronomers were able to adapt the ideas developed in studying atomic bomb explosions to their own interests. During the adiabatic expansion phase, the expansion of the remnant slows down, the lost kinetic energy of expansion being used to heat the interstellar material swept up by the shock wave. Phase 2 continues only up to a time when the total energy radiated becomes a significant fraction of the total energy of the remnant. Beyond this point, the adiabatic assumption is no longer valid—the remnant has entered Phase 3.

When Phase 3 is reached—the radiative expansion phase—losses caused by the radiation of the remnant's energy are high, and the expansion speed slows rapidly. Material swept up by the expanding shock quickly cools and forms thin, dense sheets or filaments that radiate in the optical and ultraviolet ranges. The remnant is now a rather thin shell ploughing through the interstellar medium. Eventually, the expansion of the remnant is slowed to the point where it is moving no more rapidly than the clouds of randomly moving material lying between the stars. The remnant gradually loses its identity and merges with the interstellar medium. It has entered Phase 4—the extinction phase.

APPENDIX 16: Supernova Progenitor Masses from Supernova Birthrates

Accepting an average interval between Galactic supernovae of about 20 to 30 years, and assuming that Type I and Type II events occur in about equal numbers, then we can take an average interval of about 50 years between events of the same type. While it is agreed that it is the more massive stars that undergo core collapse to be seen as Type II supernovae, it is unlikely that very massive stars, greater than about 20 solar masses, could produce the Type II's—their massive cores would collapse to become black holes. Of course it is unlikely that a black hole so formed could swallow the remainder of the stellar outer envelope without burping violently. However, such an outburst would not be expected to be a normal Type II, and a class of silent supernovae (such as was proposed for Cassiopeia A) has been suggested for the death throes of a truly massive star that might already have undergone substantial mass loss during its short, turbulent life. If Type II events occur on average every 50 years, and knowing the approximate numbers of stars in various mass ranges, it is possible to deduce that they must be formed by *all* stars in the mass range of about 8 to 20 solar masses that undergo core collapse to form neutron stars with the loss of their outer envelope (containing about 80 percent of the star's mass). In this scenario, *all* Type II events would produce a neutron star, which if an active pulsar could sustain a plerionic remnant. This leaves the progenitors of the Type I supernovae as being less massive than 8 solar masses. Since there are vastly more such small stars than large ones, only about one in every thousand of the smaller stars would need to explode as a Type I supernova to produce one such outburst every 50 years or so on average. However, we are still some way from deciding what produces the Type I events.

APPENDIX 17: Radioactivity and Type I Supernovae

Radioactivity is the process in which particles are emitted from an unstable nucleus in an attempt to achieve stability. Three types of radioactivity are relevant: alpha emission involves a particle consisting of two protons and two neutrons (equivalent to a helium nucleus); in beta emission a neutron in the nucleus breaks up into a proton (retained in the nucleus) and an ejected electron; and, finally, positive beta emission results in the emission of a positron (a particle of identical mass to an electron but carrying a positive electric charge) and the conversion of a proton to a neutron in the nucleus. In Type I supernovae the radioactive process is believed to involve up to one solar mass of a radioactive isotope of nickel decaying by positive beta emission to cobalt, and subsequently by the same process, but more slowly, to iron. Certainly the light curve and spectral properties of Type I supernovae can be adequately modeled by such a radioactive decay process, and it is then necessary to identify what type of star could explode with the production of up to a solar mass of radioactive nickel. An important implication of the thermonuclear explosions that could produce sufficient nickel to explain the light curve is that they would completely disrupt the star, leaving no compact remnant. This certainly fits the observational evidence that the remnants of the 1006, Tycho's, and Kepler's supernovae, all believed to be of Type I, show no evidence of a central neutron star. It has already been suggested that Type I events could be triggered by accretion onto the surface of a white dwarf. The requirement for significant nickel production limits the type of accreting white dwarf that could be involved—one possibility is a white dwarf of degenerate carbon and oxygen; another possibility is a helium white dwarf. The earlier loss of the outer hydrogen envelope of either type of object would explain the absence of hydrogen lines in the spectrum of Type I supernovae. There is no shortage of potential white-dwarf candidates for Type I explosions—the main requirement is that they be driven to sufficiently high mass by accretion, while the arrival rate and composition of the accreted gas plays an important role in the explosion process. However, it has not yet been possible to identify a single unique progenitor class for Type I supernovae.

APPENDIX 18: Cosmic Rays

Cosmic rays are particles that reach the Earth's upper atmosphere after traveling through space at almost the velocity of light. Because the Earth's atmosphere acts as a screen to primary cosmic rays, it is necessary to study them with detectors placed on mountaintops, flown on balloons, or launched on rockets or satellites.

The bulk of the cosmic rays (some 90 percent) are protons, while the rest are electrons, alpha particles (helium nuclei), or other nuclei. All carry electrical charge and are therefore subjected to deflecting forces in passing through magnetic fields. Low-energy cosmic rays are therefore deflected by the Earth's magnetic field before reaching the upper atmosphere. Farther out they would have had to run the gauntlet of the interplanetary and interstellar magnetic fields, the latter confining the particles to the disk of the Galaxy. Cosmic rays seem to be a background feature of the interstellar medium of the Galaxy, and their total energy has been estimated to be enough to play a significant role in the heating and ionization of the interstellar medium.

Supernovae are now recognized as the probable major source of cosmic rays, just as Zwicky and Baade had suggested. But how do the particles attain velocities close to that of light? The supernova shock wave blasting its way out through a star is one possible accelerator. As it reaches the outer layers of relatively low density, its energy is transferred to such a small total mass that individual particles are accelerated to extreme velocities. An additional acceleration mechanism is provided by fast-moving knots of magnetic field in supernova remnants. Finally, pulsars may also contribute to the cosmic-ray flux.

GLOSSARY OF WORDS USED FREQUENTLY IN TEXT

Absolute magnitude—the visual magnitude a star would have if placed at a standard distance of 10 parsec (32.6 light years).

Absorption—reduced intensity of radiation as it travels through some medium.

Absorption spectrum—dark (absorption) lines superimposed on a continuous spectrum.

Accretion—the process by which an astronomical body gains mass from its surrounds.

Accretion disk—the disk of accreting matter surrounding a star.

Alpha particle—the nucleus of a helium atom, containing two protons plus two neutrons. Emitted from certain radioactive elements.

Alpha process—synthesis of certain elements by fusion of alpha particles.

Apparent magnitude—a measure of the visual brightness of an astronomical body observed from Earth.

Astronomy—the science describing the nature and evolution of celestial bodies.

Astrophysics—the use of the methods and principles of physics for the study of astronomical systems.

Atom—the smallest particle of an element that exhibits all the properties of that element.

Barred spiral—a galaxy with a bright bar-like structure at its center from which spiral arms trail.

Beta particle—an electron emitted by certain radioactive elements.

"Big Bang" theory—the cosmological theory positing a single moment of creation.

Binary star—two stars, held by gravity, orbiting each other.

Black hole—an object that has suffered gravitational collapse and whose gravitational field is so intense that even light cannot escape from it.

Cepheid variable—a type of giant variable star whose light varies periodically in intensity.

Clusters of stars—groups of stars with a common origin, bound together gravitationally.

Continuous spectrum—the spectrum of light that appears as a continuous blend of colors.

Core—the central volume of a star.

Cosmic rays—energetic particles, possibly originating in supernovae, that pervade the Galaxy.

Cosmology—the study of the origin, evolution, and ultimate fate of the universe.

Degenerate gas—a state of matter in which the availability of energy states dictates the behavior of the matter.

Doppler effect—the apparent change in frequency of sound or light caused by relative motion of source and observer.

Electromagnetic radiation—propagating waves of electric and magnetic field; includes light, radio waves, infrared and ultraviolet radiation, X- and gamma rays.

Electromagnetic spectrum—the complete range of electromagnetic radiations.

Electron—a fundamental subatomic particle of small mass and negative electric charge—orbit central nucleus of an atom.

Element—a basic substance that cannot be reduced to simpler substances by chemical means.

Emission line—a discrete line feature in a spectrum.

Emission spectrum—a spectrum displaying emission lines.

Escape velocity—the velocity an object must have to escape the gravitational attraction of an astronomical body.

Extragalactic—lying outside our own galaxy, the Milky Way.

Fission—the break-up of massive atomic nucleii, with the release of energy.

Fusion—the fusing together of light atomic nuclei, with the release of energy.

Galaxy—a large conglomerate of billions of stars—may be spiral, elliptical, or irregular. With first letter capitalized, refers to Milky Way.

Giant star—a luminous evolved star of large size.

Gravity—the universal attractive force that acts among all matter.

Guest star—Chinese term embracing novae, supernovae, comets, etc.

Helium flash—the explosive-like ignition of helium burning in a star's core.

Hubble's law—the velocity–distance (V–d) relationship for galaxies. $v = Hd$ H is the "Hubble constant."

Index Catalog (IC)—a supplement to the New General Catalog (NGC) of nebulae.

Infrared radiation—electromagnetic radiation of wavelength longer than visible red light.

Interstellar medium—the gas and dust among the stars.

Ion—an atom which has either an excess or deficiency of electrons.

Irregular galaxy—a galaxy of irregular shape; that is, neither of spiral or elliptical form.

Isotopes—forms of an element, in which atoms of different isotopes contain an identical number of protons, but different numbers of neutrons.

Law of red shift—see Hubble's law.

Light curve—a plot of varying intensity with time of an astronomical object.

Light year—the distance light travels through space in one year.

Local group—the Milky Way plus the nearby galaxies which appear to be gravitationally linked.

Luminosity—the rate at which an astronomical body radiates electromagnetic energy into space.

Magnitude—a measure of how bright an astronomical body is (see absolute magnitude and apparent magnitude.)

Messier catalog—a catalog of nebulae prepared in the 18th century by the French comet hunter Charles Messier.

Milky Way—the nebulous band, made up of the light from a myriad of stars, stretching across the sky—the name for our parent galaxy (alternative to *the Galaxy*).

Nebula—a luminous cloud of interstellar gas and dust.

Neutrino—a minute particle of near-zero or zero mass and no electric charge; a product of certain nuclear reactions.

Neutron—a subatomic particle without electric charge; with protons, a basic building block of atomic nuclei.

Neutron star—a small extremely dense star composed almost entirely of neutrons—formed by the gravitational collapse of the core of a massive star—may be observed as a pulsar.

New General Catalog (NGC)—a catalog of star clusters, nebulae, and galaxies—successor to the Messier catalog.

Nova—the violent explosion of a star caused by thermonuclear runaway, so that it brightens suddenly then fades over days to weeks.

Nucleus—the dense central core of an atom; composed of neutrons and protons.

Obscuration—the absorption of starlight by interstellar dust.

Parsec—an astronomical unit of distance.

Period-luminosity relation—the relationship between the absolute magnitude and period of light variation of Cepheid stars.

Photon—a discrete packet of electromagnetic radiation.

Photosphere—the visible surface of a star.

Planetary nebula—the shell of expanding gas around an evolved star or post-nova star.

Polarized light—the directional alignment of the electric component of an electromagnetic wave.

Positron—a subatomic particle of equal mass to the electron but positive electric charge.

Proton—a subatomic particle of positive electric charge; with neutrons, a basic building block of atomic nuclei.

Pulsar—a rapidly rotating neutron star, identified by regular pulsating radio emission.

Quasars—compact, extremely energetic extragalactic objects—possibly the energetic nuclei of proto-galaxies.

Radio telescope—a telescope designed for receiving cosmic radio emissions.

Radioactive decay—the decomposition of certain radioactive elements, when energetic particles (alpha or beta particles) are emitted by an unstable atomic nucleus.

Red giant—a large, cool, very luminous evolved star.

Red shift—the Doppler shift of the radiation from distant galaxies, attributed to expansion of the universe.

Schmidt telescope—a reflecting telescope with very wide field of view, used principally for astronomical surveys.

Seyfert galaxy—a spiral galaxy, showing evidence of violent upheaval within its nucleus.

Spectrograph—an instrument used to obtain the spectrum of a source of light.

Spectrum—the luminous band varying from violet through to red produced when light is dispersed into its component colors.

Spiral galaxy—a galaxy with spiral arms trailing from a bright central nucleus.

Supernova—the violent self-destruction of certain types of stars that have reached the end of their normal evolution.

Supernova remnant—the debris and swept-up interstellar material from a supernova explosion.

Synchrotron radiation—the radiation emitted by charged particles spiraling in a magnetic field.

Ultraviolet radiation—electromagnetic radiation of wavelength shorter than visible violet light.

Variable star—a star whose brightness varies with time.

White dwarf—a star, of mass less than the Chandrasekhar

limit (1.4 solar masses), that has expended all its available nuclear energy.

Wolf-Rayet stars—very hot, highly evolved stars that shed their outer layers to space.

X-rays—electromagnetic radiation of wavelength shorter than ultraviolet light.

INDEX